Female Choices

Also by Meredith F. Small

Female Primates: Studies by Women Primatologists (editor)

FEMALE CHOICES

Sexual Behavior
of Female Primates

Meredith F. Small

Drawings by Andrea S. Perkins

Cornell University Press
Ithaca and London

Copyright © 1993 by Cornell University

Drawings copyright © 1993 by Andrea S. Perkins

First published 1993 by Cornell University Press.

Printed in the United States of America.
Color photographs printed in Hong Kong.

⊗ The paper in this book meets the minimum requirements of the American National Standard for Information Sciences— Permanence of Paper for Printed Library Materials, ANSI Z39.48-1984.

Library of Congress Cataloging-in-Publication Data

Small, Meredith F.
Female choices : sexual behavior of female primates / Meredith F. Small ; drawings by Andrea S. Perkins.
p.   cm.
Includes bibliographical references and index.
ISBN 0-8014-2654-5
1. Primates–Behavior.   2. Females.   3. Sexual behavior in animals.   I. Title.
QL737.P9S595   1993
599'.80456–dc20
92-56785

To my sisterhood—Ann, Audrey, Becky, Candace, Claudia, Carol, Dede, Didi, Diane, Jeanne, Kaajia, Kristen, Pat, Sandy, Ute . . .

And to my sisters—Krysia and Andrea

Contents

Contents • viii

Acknowledgments

MY KNOWLEDGE of sexual selection theory, female mating strategies, and primate behavior in general was initiated and nurtured by many scholars of animal behavior, anthropology, and primatology. To them I owe a great debt. They gave me what any aspiring academic or writer might want—the atmosphere in which to expound on crazy ideas with only gentle (but accurate) criticism that kept my feet securely glued to the ground. I thank past and present members of my University of California, Davis primate cohort, my academic family: Leo Berenstain, Michael Ghiglieri, Sarah Blaffer Hrdy, Lynne Isbell, Maureen Libett, Fred Lorey, John Mitani, Ryne Palombit, Peter Rodman, Linda Scott, David Glenn Smith—all scholars, teachers, and friends. I must single out Ryne Palombit for a special thanks. Without him, this book would never have been conceived. We have spent many hours discussing female choice, and Ryne's insights and his insistence on clarity of thought have shaped my approach to female choice more than he might realize.

Other primatologists also influenced this book. They directly affected my work with comments on manuscripts, encouraging words, or long talks about female behavior. I thank especially John Berard, Richard Byrne, Robin Dunbar, Alison Jolly, Jane Lancaster, James McKenna, Ute van den Bergh, and Patricia Wright. I am particularly grateful to Ute van den Bergh and Jane Lancaster for reading and commenting on the entire manuscript. Frans de Waal and Elizabeth Wardell were kind enough to lend photographs to illustrate important points in the book.

I and other female scholars are especially indebted to Sarah Blaffer Hrdy, Alison Jolly, Jane Lancaster, and other women primatologists who made us wake up to what female monkeys and apes are really doing, not just what they "should" be doing. This book only follows in their footsteps.

During the spring of 1991, I conducted an undergraduate seminar on female choice in primates at Cornell University. The fifteen students in that class independently came to many of the conclusions in this book. Those weekly discussions helped hone my own ideas and gave me confidence that I was on the right track.

I owe a special debt to science writers Nathanial Comfort, Bruce Lewenstein, Mike May, John Pfeiffer, Wallace Ravven, and James Schreeve. They were the first to look at my nonacademic work and to encourage my detour into popular science writing. My "writer friends," Diane Ackerman, Jim Gould, and Jeanne Mackin, encouraged me with their constant faith. I also thank Barry Lopez and Robert Finch, who, unbeknownst to them, taught me to write not just about what animals do, but how I feel about them.

Special thanks go to my sister, and the illustrator of this book, Andrea Small Perkins. Not only are her drawings beautiful, they also capture the essence of each chapter. Some might think that her ability to reflect my words so perfectly, and with a sense of humor, comes from our having half our genes in common. But I realized soon after I saw the first drawing that this is Andrea's special talent—she can digest a writer's words and produce those words in a picture; I only feel lucky that she offered to do this for me. In addition to creating the inspired drawings, Andrea was also the first person to read the manuscript. In that capacity, she served as a naive reader and as first editor. Most important to me, she was kind enough to impart encouraging and complimentary words during the last few months of writing.

Other friends and members of my family, those who know nothing about monkeys but much about me, have provided a personal foundation. Ann and Don Jereb have the large telephone bills to prove it. Becky Rolfs, as always, provided levity. My sisters, Krysia Bruck and Andrea Perkins, my brother, Charles Small, and my parents gave familial reassurance about my writing by their eagerness to read it all. And I thank Tim Merrick, my consort partner, who now knows more about female choice than he ever might have wished.

The beginnings of my thoughts on female choice first appeared as a feature article, "Girls Just Wanna Have Choice," in *Science Illustrated* (June 1988: 28–32). Four years later, those ideas had become fine tuned and appeared as an article for *American Scientist*, "Female Choice in Mating" (80 [1992]: 142–151), which is essentially

Chapter 4 of this book. I write about our chimpanzee cousins, the bonobos, in both Chapters 5 and 6, and this material appeared earlier in *Discover* as an article on bonobo sexual behavior, "What's Love Got to Do with It?" (13 [1992]: 46–51). Another article for *Discover*, "Sperm Wars" (12 [1991]: 48–53), helped me understand the details of human conception, and some of that information is contained in Chapter 5. I wrote about primate social intelligence for *The Sciences*, "Political Animal" (March/April 1990: 37–42), and mention this concept when discussing how smart primates are in Chapter 2. I also use Barbary macaques extensively as examples and have written about them in the popular press. In particular, Chapter 3 notes the social climb of a female monkey up the hierarchy, a story I told previously in *Natural History* ("Ms. Monkey," January 1989: 10–11). I also write about baby macaques in Chapter 3, and this work was first published in *Bulletin*, Field Museum of Natural History, as "Babysitting and Daycare among the Barbary Macaques" (60 [1989]: 24–28). I thank the editors of these magazines for permission to use this material here.

Finally, I must explain to fellow primatologists how this book was constructed. I have tried to be faithful to the current information on primate female behavior. Those who have written for a non-academic audience will appreciate my attempt to present the science accurately without boring readers with endless numbers, charts, and graphs. I walked this tightrope by using certain species as examples, very often macaques because I know them best, and then referencing other studies rather than explaining them all in detail. The information on primate behavior from field and laboratory studies has grown exponentially over the past fifteen years; I focused my reading on studies that concern female sexual behavior and mate choice. Sometimes I may have failed to mention, or reference, an article written by one of my colleagues. This omission was born not of malice, sloppy scholarship, or an attempt to skew the available data in a particular direction. I may have felt a reference was not relevant to the perspective here, or one may simply have skipped my notice. I apologize if I overlooked anyone, any species, or any confirming or contradictory data, and I hope that the articles in question will be brought to my attention in time for *Female Choices II*.

M. F. S.

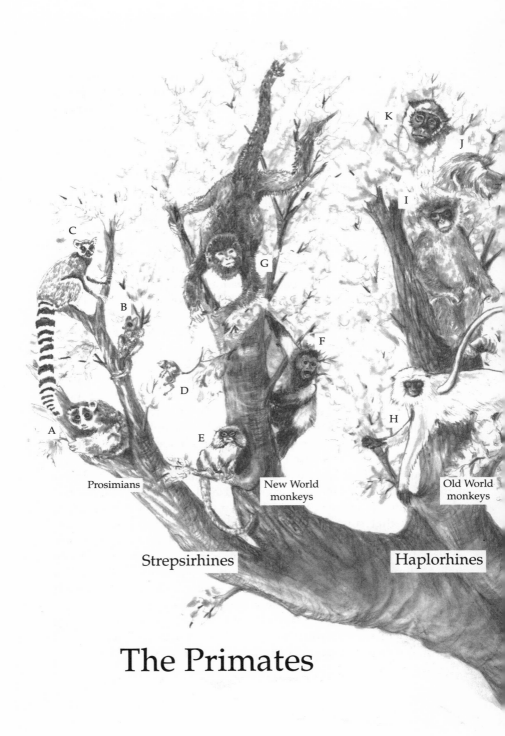

C

B

A

Prosimians

Strepsirhines

G

D

E

New World
monkeys

F

K

J

I

H

Old World
monkeys

Haplorhines

The Primates

M

O

N

P

Apes

L

A, Loris

B, Galago

C, Ringtailed lemur

D, Tarsier

E, Marmoset

F, Capuchin

G, Howler monkey

H, Hanuman langur

I, Barbary macaque

J, Baboon

K, Vervet monkey

L, Gibbon

M, Orangutan

N, Gorilla

O, Chimpanzee

P, Human

Female Choices

Introduction

AS AN ANTHROPOLOGIST who studies nonhuman primates, I might have chosen to focus on any set of behaviors. I picked sexual activity, mating, and mate choice because they are the basic stuff of evolution. Evolution works though the two processes of differential survival and differential fertility. Differential survival means that some individuals live, and live longer, than others. The specific traits and behaviors that help an animal find the right food and avoid danger have evolved because they contribute to its survival. The way it lives—eats, moves about, and avoids predators—is vital. This book, however, is concerned more with differential fertility— the fact that some individuals reproduce more than others. Any animal must live long enough to pass on genes to the next generation if it is to have a chance at reproductive success. Survival tactics mean nothing, however, if the animal never mates and reproduces. Therefore, the nitty-gritty of evolution is reproduction, especially who reproduces the most. My focus on sexual strategies in this book grew from my interest in how all primates go about their mating business and in what determines the winners and losers in the game of reproductive success. So this book is more about sexual strategies than sexual activities in general.

I also chose to focus on females rather than males. The reason is obvious—I'm a female looking for answers about her own behavior. All scientists, if they're honest with themselves, study something that has relevance to their lives. In my case, I want to know what makes me and other female primates tick, understand our similarities and our differences, and discover what our shared history means to the evolution of human female behavior. We female primates share an innate urge to have sex and reproduce; we share enlarged body size, long periods of our infants' dependency, the requirements of lactation, and slow moving life histories with long periods at each stage. We're literally sisters under the skin, right down to shared fragments of DNA. For example, I do a perfect imitation of a female macaque distress call, and not just because I'm good at imitating animal sounds: the emotion of distress is as real to me as it is to that macaque in trouble. It's also no wonder that when I see a female gorilla clutching her infant I feel empathy. Or when a female baboon defends her sisters, or forms an alliance with a friend to run off an obnoxious male, the scene looks familiar. Because of a shared genetic history, much of what human females do is paralleled in the behavior of our nonhuman primate cousins—prosimians, monkeys, and apes.

There's something else all primate females share—the unpleasant experience of being talked about behind our backs. Orgasms, sexual pleasure, mate choice, sexual interest are components of female sexuality that scientists and writers have either discarded or overemphasized according to social fashion. Only recently have researchers asked human females directly about their sexuality or spent time watching what nonhuman female primates really do during mating. This book is an attempt to put views of female sexuality and mating behavior on the right track. I don't answer all the questions about female mating, the data just aren't there. But at least this book does demonstrates that females are active sexual participants, creatures making decisions about their own reproductive futures. We females were once considered shy and passive sexual beings, but science has finally come to acknowledge what each of us already knows: we enjoy sex. This simple fact has been punished by society, ignored by men, and discouraged by families. Even the word "promiscuous" is used almost exclusively in reference to women who seek sexual gratification with several partners, with the implication that sexual enjoyment by women is pathological. When males play the field they are sometimes called satyrs, and the image is one of harmless randy creatures having fun. Science has given theoretical support to this male-female sexual dichotomy, telling us that men "should" have more sex, and with more partners.

Such behavior improves their chances of passing on genes to the next generation because they are spreading sperm among various women. Women, on the other hand, "should" be passive, selective, and stick with one good man, in order to protect their investment in their more limited breeding capability. But contrary to what theory suggests, female primates just keep on being highly sexual. They even dare to mate with several males.

Some might ask, if I want to write about human female sexuality, why not just write about humans and their recent history? The problem is that we actually know very little about human sexual strategies. Why one person chooses another as a sexual partner or lifetime mate mystifies us. We can speculate on how culture imposes all sorts of rules for choosing a partner and how these rules make economic, religious, or social sense. But still there are missing pieces of the puzzle, and I maintain that we can discover those pieces, in part, by looking at what other primates do. If we want to know why someone acts in a certain way, we look at her family history. In a broader sense, this book is doing the same thing—looking at the very large picture of human female sexual behavior by taking into account what all primate females do.

Anthropologists like myself try to figure out why present-day humans are the way they are. To do this, some anthropologists look at people in every culture, consider what our primate cousins are doing, and project into the past. After all, we weren't beamed down here standing on two legs, living in villages. We began at the split between apes and human about 8 million years ago, when human and chimpanzee ancestors shared a common relative. This ancestor wasn't exactly like a chimpanzee of today, nor was it particularly human. The chimp line diverged from this common ancestor and later it split, producing chimpanzees and bonobos. The human lineage moved on from that common ape-human ancestor. We evolved over those many millions of years into what we are today. Anthropologists must use a patchwork approach to piecing together our past—examining some fossil anatomy here, archaeological remains there, and what other primates are doing added in. The point is to construct a pattern of human nature. Perhaps the most difficult step in that reconstruction is the nature of human sexuality and mating. For every scenario builder, there's a new picture of human sexual evolution, and with each new piece of information about our past, a particular scenario must be poked, prodded, and reshaped. This ongoing process has left us with few answers, but many intriguing speculations about why human females mate the way they do, and why the human family and mating system evolved.

As an anthropologist, I use nonhuman primates to delve into the biological roots of human behavior. This might scare some readers. We humans, especially in Western culture, like to believe we have free will, that everything we do is a conscious decision consistent with our own cultural, psychological, or emotional makeup. The perspective of this book is quite different. When I take the evolutionary view, one that treats us as any animal species and ties all primates together, I'm proposing that there is a general human nature which is part of a broader primate nature. More important, I suggest that this nature is a driving force underlying much of our individual behavior.

The only way to grab at the biological component, we primatologists believe, is to compare ourselves with other primates. Some, especially social anthropologists, maintain that humans are so different, so special that biology has no impact on what we do. But as John Mitani, a primatologist, points out, this is a comparative statement (Small 1993). With whom is the comparison to be made, we might ask? Obviously, our living primate relatives. If we then place human attributes next to the behavior of other primates, what do we get? Surprisingly, the comparison starts to break down. Ever since Jane Goodall began reporting on the behavior of chimpanzees, we've seen the wall between human and nonhuman primates tumble. We once thought humans unique because they were the only species to regularly use tools. But chimpanzees also make and use tools on a regular basis. Complex communication, too, was thought to be an indestructible barrier that separated humans from other animals. But we've since learned that primates, and other animals, communicate with intricate sophistication about things, feelings, and what they know. Even our social and kinship networks pale in comparison with those of other primates. Macaques, for example, live in a social whirl that rivals any soap opera. Their ability to recognize one another from calls alone is certainly better than the average human's ability to recognize the scream of a friend at a distance. In fact, attachment to kin and personal loyalty to friends among macaques are qualities humans could emulate.

Although humans may differ in degree in the ways they utilize these behaviors and abilities, they don't differ in kind. And this statement may also be false. Who's to say that chimpanzee society is less complex than a human village? What makes typing the keys of a computer more complex than fashioning a fishing rod for termites? We shut ourselves off from understanding who we really are when we consider ourselves unique—more complex, and by implication better, than other primates. This need to be "special" gets in the way of discovering our place in Nature.

The human and nonhuman female primates in this book are highly sexual individuals who enjoy sexual activities for their own sake and make their reproductive choices according to what's important to them. They compromise when need be, but they play the game of reproductive success with a skill that matches that of their male partners. These females are not the passive, shy creatures of Victorian times, but individuals who often decide the course of reproductive events for a couple, influence the genetic future of a group, and can affect the overall mating and social system of a species.

The book starts off on a sexual note in Chapter 1. But this chapter has nothing to do with the higher pleasures of sex. Instead, I attempt to explain why we have sexual reproduction at all. One of the major mysteries of nature is that although most creatures on earth reproduce sexually, no one really knows why. Asexual reproduction does fine for some organisms, but evolution has also opted for sexually reproducing creatures that need to mate with another individual to pass on genes. Understanding this basic step in evolution is the introduction to the larger issue of the specific female role in sexual reproduction. Chapter 2 introduces the primate order. Everyone knows that chimpanzees are primates, but there are 199 other primate species that have something to tell us about our background. Much of what we know about the evolution of human behavior comes from work on macaques, baboons, and other primates. They are introduced here as a cast of characters that reappear in later chapters. Chapter 3 is another introductory chapter, but it focuses specifically on female primates. I follow the life course of one macaque and explain what her life is like from birth to old age. Here the reader should see similarities to and differences from a human female's life. Chapters 2 and 3, I hope, establish a bond between the reader and the other animals he or she is reading about. With Chapter 4 begins the journey exploring female sexuality. Female choice, a sexual strategy, is described in historical and theoretical terms. We see females tossed about by the social and scientific milieu as passive or wanton, choosy or careless. The theory of female choice as part of Darwin's sexual selection theory has undergone a revolution in recent times, but the change in attitude may not be quite enough to accommodate the mating behavior of female primates. The contrast between what female primates "ought" to do and what they really do becomes clear in Chapter 5. A picture of a sexually assertive female emerges in these pages as I discuss conception, orgasm, and homosexuality in female primates. Chapter 6 is the core of the book. Based on recent work on mating in primates, it evaluates how females make their mate selections. Surprisingly, preferences for novel, unfamiliar males stands out as the major feature of female mating strategies. So-called promiscuity is also a part of primate female sexuality. In some ways, Chapter 7, the final chapter, is the apex of this book because it focuses on humans. But I hesitate to make that claim because it implies that humans are on a higher plane than other primates. But still, humans like best to read about themselves, so I suspect that this is the chapter most readers have been waiting for.

As the title suggests, *Female Choices* is a book about the choices made by, and available to, female primates. But these are not choices in the usual daily sense, such as which earrings to wear tomorrow or which road to take to work. The female choices in this book are ones of mating and reproduction—the decisions with larger consequences which shape not only one's day but one's life. And none of these choices is made easily because each is constrained by female biology and the drive to improve one's reproductive success. In that sense, I am concerned not only with "female choice," the evolutionary concept most often aligned with sexual selection, but also with the other mating choices that females make. Those who make the best mating decisions, consciously or unconsciously, pass on genes and are well represented down the ages. Those who make bad evolutionary decisions steer their genes into oblivion. There is no morality here, no right and wrong, but female decisions about mating, and female choices for and against particular partners, do mold the primate world we live in.

CHAPTER 1

==================

Biological Warfare

A FEW YEARS BACK, I found myself yelling at the television set. The object of my tirade was talk-show-host Phil Donahue. There was the usual panel of men and women discussing their difficult relationships, which are always fascinating, but this time Phil had wandered in to evolutionary territory. "I don't get it," he said. "If the whole idea is to reproduce the species, why are men and women so angry with each other?" I was shouting at this statement because Phil had it all wrong. He just didn't understand that a war between men and women is exactly what evolution predicts. It is true that the tension between men and women today is almost palpable. We often see angry discussions on TV between disappointed women and defensive men, like the show I was watching. Cliques form at cocktail parties—men in one corner and women in another—all whispering the same sweeping generalizations about the opposite sex. Men are so thoughtless, women so demanding. Is this face-off simply a result of unrequited expectations, lack of communication, or maybe just plain different views on how human relationships should operate? Probably all three, but there's an additional piece in the male-female puzzle, a piece so primal, so fundamental, that it's the foundation of the puzzle itself. Our inescapable biological heritage, in which females produce infants and males do not, naturally pits females and males against each other. And this difference is as true for squirrels, lions, and monkeys as it is for people. The battle of the sexes can be explained, at its deepest level, as a war of different mating strategies.

Some maintain that people are supposed to act cooperatively to make more of our own kind and that collaborating on parenting is "natural" for humans. But our species, *Homo sapiens*, like any other species, has no lofty purpose, no unifying push to make sure that its lineage goes on forever. It's a nice idea, but evolution simply doesn't work that way. Evolution by natural selection operates not at the level of the species but at the level of the individual. The evolutionary goal of each individual is to pass on genetic material to the next generation, regardless of the interests of others. Evolution thus calls for an ultimate selfish strategy of looking out for one's own genetic future. But how the two sexes pass on their genes differs so dramatically that males and females of every species end up fighting each other all the way into their reproductive futures.

The problem is this: all male animals continually produce many small gametes, sperm, and they can potentially father hundreds of offspring in a lifetime. Females, with their slowly maturing eggs produced only at intervals, have fewer chances to pass on genes. In addition, only females become pregnant, and the major burden of parenting falls on she who lactates. But the more-tortoise-than-hare strategy of females usually results in at least some reproductive success over a lifetime, while males, who hastily spread their sperm, often end up with nothing. They may mate with several females but never actually conceive because their hasty copulations are ill-timed. Males and females are caught in a conflict because first and foremost they need each other to make babies and pass on genes, but they use radically different strategies to attain that goal. They eventually compromise, but it's a tenuous truce.

These simple truths, ones humans and all mammals share, have significant evolutionary consequences. They dictate how individuals choose each other as partners, the patterns of mating and association, and the way individuals raise their offspring to sexual maturity. The long shadow of reproductive biology also influences female-male social interactions in all species and determines how the sexes treat each other day to day. Compelled by the urge to pass on genetic material to the next generation, the sexes must often cooperate in mating and parenting, but each sex cooperates only under duress because females and males operate under different reproductive rules set down in opposing directions eons ago. Like an open wound that never heals, the conflict between males and females will never be resolved because the evolutionary interests of the two sexes are forever locked in opposition. Each sex needs the other. There's no right or wrong here, no sex better than the other, just two types of individuals trying to win in the game of reproductive success. The ground rules of the battle include cooperation, conflict, and exploitation, and both sexes use these tactics equally.

This book views the war mostly from the female side. Female animals, including humans, have been portrayed in literature, history, and science as mysterious, coy, or manipulative. And nowhere is the female role, or the male role for that matter, more ambiguous than in stories of mating. While males have been stereotyped as sexually aggressive and powerful, females have been portrayed as passive or, alternatively, conniving sexual partners. Views of female behavior have also echoed societal norms for behavior, endorsing Victorian virgins or liberated feminists. Recently, based on a growing body of information collected over the past two decades by scientists in the field and laboratory, a new picture of female life histories is being drawn. At least for nonhuman animals, females emerge as strategizing creatures acting in ways that differ from those of males. But the most important lesson gained from the past few decades of animal research is that reproductive strategies of female animals, especially their mating behavior, is tuned to the particular needs of females. What we might hope to gain by looking at one sex is a clearer picture of how certain tactics appeared and why they aid in passing genes to the next generation. In this way we can make some sense of the natural world and begin to comprehend why humans, and other animals, resolve their biological differences long enough to mate and bring up offspring.

Shadows of Love

The only way for an animal to pass on genes to the next generation is to mate, conceive, and bring forth healthy infants. It follows that patterns of mating should be subject to natural selection. This dance of love and mating is the evolutionary heartbeat of behavior—without mating there would be no representation in future generations. We should expect evolution to act swiftly in favoring some patterns over others. Those who choose good mates, make healthy babies, and help these infants reach maturity will be the winners in terms of reproductive success. It's just a simple matter of numbers; those who employ the best tactics for mating and reproduction pass on more genes. How they go about mating, whom they choose to mate with, and why, are all factors that have been shaped by natural selection. Specifically, this process is the evolution of sexual behavior and mating strategies.

To accept this fact, one has to understand the mechanics of evolution. Through time, the genetic makeup of any species changes. It does so because some individuals live and some die, and the genes of some individuals are passed on while those of others are not. This difference in mortality and fertility determines what the next generation of that species will look like. Those who can best deal with their environment, avoid predators, get enough nutrition, find mates, and keep their infants out of harm's way will pass on their particular pool of genes. Any traits or behaviors that aid an individual through these tough life stages will be passed on, selected for. And traits that get in the way, traits in individuals that don't live or reproduce, will be selected against. Natural selection, the major force of evolution, selects for and against individuals who carry genes that might or might not be represented in the next generation of that species. This is how species evolve, change, and adapt to particular environments.

It's easiest to see how evolution works when we consider how organisms *look*. For example, some species of birds have pointed beaks because their ancestors were able to exploit a food resource that required drilling into tree bark. Those individuals with the most-pointed beaks ate more, lived longer, and had more baby birds than (ones with pointed beaks) birds with rounded beaks. Pointed beaks were selected for. We can use this simple example to understand the appearance of particular features such as body shape or special traits that allow organisms to exploit certain features of an environment. But the effect of evolution on *behavior* is much more difficult to unravel. Behavior is not often a do-this-or-that phenomenon. Behaviors aren't really discrete features of an individual. Behaviors happen as part of other actions, they are patterns rather than single events. But behavior, be it sunbathing by lizards, courtship displays of birds, or cooperative hunting by chimpanzees, has also been molded through natural selection. Patterns of behavior, just like morphological features, are subject to the rules of selection. An animal doesn't stay alive and make offspring only because it looks a particular way; how that animal acts is also important to its survival and reproduction.

The clearest example is an animal's ability to spot and run from predators. Those who are sharp of eye and fleet of foot have a behavior pattern that has been favored by natural selection over generations. The behaviors involved—vigilance, then flight—were also selected over time because those animals who didn't look out and run away were eaten. Their genetic line ended with the first bite. Thus the combined pattern that includes keeping watch and reacting to moves in the bushes evolved in the same way that good eyesight for spotting the predator and good leg muscles to carry the potential victim away evolved.

In the lives of males and females, the most primitive behavioral compromise occurs when individuals meet and mate. For a new individual to be formed, sperm must come in contact with an egg, and when sexual reproduction requires internal fertilization, the meeting is quite complex. Potential partners have to find each other and move into a position whereby the sperm can be deposited within the female reproductive tract. She must be interested and he must be able to deliver. For birds the compromise may mean meeting aloft in an aerial dance of sperm transfer. For monkeys it means the female must stand still and allow a much larger male to grab her, mount her back legs, and impose his weight as he copulates. And for humans, females and males move to the rhythm set by both partners.

No one knows what goes on in the minds of nonhuman animals during copulation. We know the sexual act only by how we experience it as humans. If asked, most people would say they have sex because it feels good. It's that simple—the sexual act feels fine while it's going on, and a sense of inner calm often follows orgasm, so the positive feelings last for quite a while. We humans experience the mating process as a pleasurable encounter, full of intimacy and bliss. Most people also view the sexual act as a release, but no one says what this release is all about. Is it physiological, emotional, psychological, or a combination of all three? It might be relaxing or frustrating, it might make the day worse or better, but we all have a hunger to engage in this meeting of bodies. In simple terms, humans are motivated to engage in sexual activity by the knowledge that it fulfills some pleasure gap, but the urge is one we rarely explore and barely understand. Yes, we moralize about who should or shouldn't copulate, and we gossip a lot about who's having sex with whom. We also arrive at a societal consensus full of sexual rules and try to force people to act in "appropriate" sexual ways, but the rules are constantly broken. Most of all, we really don't fathom why we need this pleasure. Like wild rabbits in a warren or deer during the rut, we move though the sexual motions without really understanding our urges or our sense of pleasure. We just do it.

And yet there's a simple, albeit profound, explanation for our sexual urgings. We copulate because we're compelled by our genes, just like all other sexually reproducing creatures, to pass on bits of ourselves to the next generation. As clinical as it may sound, the yearning for low-down erotic sex is just Nature's way of making us produce babies. The elaborate rules of culture are just trappings that cloud our biological urge to be present in the next generation. Without that urge, possibly few would reproduce at all. Objectively we recognize that sex without birth control may lead to pregnancy, but most couples in the throws of pleasure, if questioned, would probably maintain that the sexual joy of the moment was well removed from thoughts of reproduction. Tell people they copulate because their genes dictate they do so, on the off-chance an infant will be produced, and receive some odd looks. After all, the lofty sentiments of love poems rarely mention a genetic blueprint behind the verse. Falling in love never seems like the tug of some primitive leash dragging us kicking and screaming toward reproduction. But put in blunt evolutionary terms, we don't have sex because it feels good, we have sex because millions of years of evolution have hard-wired sexual desire into our brains.

This isn't to suggest that humans are preprogrammed robots with no say in their actions. The complexity of our social behavior has also been influenced by the long years of learning to act like humans in a particular culture. We learn a singular language, dress like our fellow citizens, and observe cultural practices sometimes not found in other groups. At the same time, every person behaves in ways, has certain desires and motivations, that are similar for all humans. For example, part of our human nature is the need for attachment. We bond with at least one other person either in marriage or in partnership. We live in social groups and go to a lot of trouble to maintain those groups. We make networks of friendships, alliances, coalitions, and we make our connections most easily with kin. To live in groups, we deal with interindividual competition, and sometimes we promote that very same competition. And by definition, all groups have ways to dispel strife so that life goes on (de Waal 1990). These behaviors have been selected over time because our prehuman ancestors employed these patterns and lived longer and reproduced better than did ancestors who used other strategies to get along. Deep down there's a generality to human behavior, a universal human nature, that ties us all together as one species. In the same sense, there's an even broader pattern of primate nature which encompasses all our closest relatives.

The joining of two individuals to create a third, what we call sexual reproduction between two forms called male and female, was selected over time. To understand why some individuals have higher reproductive success than others, we have to start at the beginning. And the beginning is before sexual reproduction was even invented.

The Why of Sex

Birds do it, bees do it, but no one is completely sure why they do it. From a strict and straightforward evolutionary view, an organism should just clone itself and produce offspring that contain its full complement of genes. Asexual reproducers, those who make photocopies of themselves with no help from partners, need no mates, no gender differentiation, and have no compelling reason to join with another organism that is genetically of the same ilk just to reproduce. They have, in fact, an excellent reproductive strategy—double one's genes and divide, again and again, leave a long line of descendents with the same genetic makeup, lose nothing in the reproductive shuffle. This process occurs regularly in the plant and animal kingdoms, and these species, called asexual or parthenogenetic reproducers, have been reasonably successful in populating the earth.

Parthenogenesis is a conservative and highly successful reproductive strategy. Any organism is related to itself by one, so if the individual "clones" itself, the resulting offspring will also be related by one because it has the exact same complement of genes. The asexual organism doesn't have to search for a mate or fight with others over the best potential mates. A parthenogen simply sits around fat and happy in its genetic purity, dividing into millions of daughters that look just like itself. From a purely evolutionary viewpoint, these individuals have won in the game of life, they've passed on their own genetic material unaltered (Williams 1975). This passing on of genes from one generation to another, unsullied by the genes of another individual, would be the perfect reproductive strategy if the world were a perfect place. In fact, from the parthenogenetic perspective, sexual reproduction appears to be downright disadvantageous.

First, to be sexual means having to combine your genetic material with someone else's. This is called the initial cost of sexual reproduction. You no longer pass on all your genetic material because if full complements from each parent were joined together, the resultant offspring would have double the normal number and be nonviable. Second, a sexual organism must combine his or her genes with those of another individual: one little slip in mate choice and the offspring might die from the unfavorable genes contributed by one's partner. Third, once sexual reproduction has begun, only females reproduce the offspring and there's an excess of useless males (Maynard Smith 1978). Sexually generated offspring also have a greater probability of carrying and combining deleterious mutations; that is, the parthenogenetically produced offspring is as genetically clean as the parent, but the sexually produced offspring runs the risk of matching two bad alleles, or forms of a gene, for a trait, thus bringing forth some hidden genetic anomaly, sometimes with a fatal result (Williams 1975). These "costs" of sexual reproduction, it would seem, should outweigh the advantages. If it's such a costly endeavor, why has natural selection opted for the more difficult path in most cases?

All around us are sexual species. They drop half their genetic material, join with another individual, and end up with offspring only half recognizable as themselves. Sexual reproduction, the mutual joining of genes to produce an offspring unlike either parent, has evolved in at least four phyla—animals, plants, ciliates and fungi—and is the most prominent way on earth in which organisms reproduce themselves. How did sexual reproduction evolve? What force prompted primitive cells to combine with a stranger? What moved cells to give up half their genes in a chance to combine the remaining set with another individual? What is the advantage of sexual reproduction?

Looking back on the primordial soup, scientists peer into a dark hole. There are few fossils to reveal why sexual reproduction evolved. Today, most researchers rely on primitive one-celled animals for clues about why two-party reproduction began. They also gain insight into the evolution of sexual reproduction from organisms that switch from asexual to sexual reproduction, such as some plants, parasites, and worms (Bell 1982). These may be intermediate sexual forms, switching from one type of reproduction to another in reaction to environmental conditions. For scientists, they provide windows to the moment long ago when sexual reproduction was first favored.

The story of sexual reproduction is not simply an ancient and peaceful story of two organisms joining in an egalitarian attempt to better themselves. It is instead a tale of exploitation, right from the start.

Once upon a time, more than 3 billion years ago, primitive cells now called prokaryotes ruled the earth. They were simple cells filled with fuzzy free-floating strands of DNA, the genetic information of life. The DNA wasn't even organized into packets of chromosomes yet, and the cells didn't have nuclei to hold the DNA in one place. The cell was just a blob filled with DNA. It replicated slowly but easily by splitting into two daughter clones of itself. These prokaryotes, like simple bacteria for example, lived in a world where nothing was protected from the damaging rays of ultraviolet light. Unfiltered by an effective ozone layer, UV light entered the fragile prokaryotes and caused their DNA to tie into knots, a twisting that ensured death for the cell. Many prokaryotes met their end in this way before they had a chance to clone and pass on their genetic material to the next generation of prokaryotes. But some of these primitive cells chanced upon a solution to the damage caused by UV light. In reaction to the damage, they devoured their neighbors, using perhaps half of a neighbor's DNA, or less, for a patch job (Margulis and Sagan 1985, 1988). This was no cooperative fifty-fifty proposition for both cells. Some bacteria hijacked a neighbor and incorporated some DNA, and the cell that then survived was cloned and passed down through the ages. Evolution works like that—no real goal, no purpose, just favoring those that somehow pass on their genes. In this case, the cells that were able both to repair their DNA and to clone passed on more "offspring" than those that died. The previously damaged marauding prokaryote often took half the DNA of its victim, or sometimes only a small amount. The resulting offspring was a jumble of DNA from two previous bacteria, but it still resided in the parent cell. From there, the jumbled individual cloned a few times with a face-lift. This was not the first real sexual reproduction, only the precursor, because these cells were still without real chromosomes. But because these prokaryotes were the first organisms to incorporate materials from another organism, even though they destroyed the other cell, scientists consider this process a precursor to sexual reproduction, a heralding of things to come. As the biologist Lynn Margulis says,

"The first sex had arrived, and it had come before gender, before sperm, before embryos, before even the evolution of cells with nuclei" (Margulis and Sagan 1985, p. 29).

In that same primordial soup, with bacteria eating bacteria, were a few cells that differed from the prokaryotes in several special features, and they are our true sexual ancestors. The major difference between these cells and the prokaryotes was a thin membrane that enclosed their DNA within the cell; they sported a membrane-bounded nucleus inside the cell. Much like coins stored in a wallet and dropped into a larger purse, the DNA of these cells was protected both by the nucleus wall, the surrounding cytoplasm of the cell, and by the outer wall of the cell (Margulis and Sagan 1985). Their DNA was also organized differently than that of the prokaryotes. Instead of appearing in flagellating strings of DNA, the genes were packed into sets of chromosomes. This set of genetic material was more complex than that of the prokaryotes, and it also had potential. These new cells, called eukaryotes, would eventually evolve into birds, bees, and humans, but for the time being they floated about looking for ways to pass on their genetic material. First they did it simply: they doubled and divided and doubled and divided just like their prokaryotic neighbors. They shifted their chromosomes about so that they always ended up with the same number of pairs. This system is the ancient ancestor to mitosis, the process whereby the cells in our modern bodies replicate themselves throughout the day, a within-organism act that replaces blood, skin, and other disposable cells. Mitosis also allows animals and plants to grow by adding the appropriate cells though simple doubling and division.

The next stage of evolution for the eukaryotes was somewhat different, however. Besides doubling their own cells, several of the eukaryotes began to exploit other cells. They reached out and devoured a neighbor, nucleus and all—not in response to DNA damage, some scientists think, but from a need for nutrition and water, a grasp for cytoplasm in a primordial soup lacking in foodstuffs (Margulis and Sagan 1985, 1988) This process might have produced a cannibalistic feast, one cell devouring another, but for one glitch. The predator often incorporated the victim's DNA into its own original complement while it nibbled on cytoplasmic treats. Instead of taking only some of the DNA, now packed in chromosomes, the eukaryotic cannibal actually fused equally with the other cell. The resultant new cell had

double the normal number of chromosomes, and this situation became exacerbated when the cell did its normal doubling routine and
ended up with four times the normal number of chromosomes. As a
result, most of the combined eukaryotes died (Margulis and Sagan
1988). But some of them dealt with the aberrant number—they simply
divided twice. The process of cannibalism and later division produced
new daughter cells with mixed arrangements of chromosomes from
predator and victim. These cells also did better than those that didn't
incorporate others. Those that failed to take advantage of their neighbors died from lack of nutrition and their genes were forever lost.
Thus the tendency to pull in another cell, to "mate" with another
cell, was selected.

   This repeated division after fusion was the precursor to meiosis,
the process our sex cells, our gametes, undergo to make sperm and
eggs that contain only half the normal number of chromosomes. Any
cell contains a set of chromosomes which occur in matching pairs,
called homologous chromosomes. The number of pairs is specific to
a species, and for a cell to operate normally it must have both homo-
logs of each chromosome. If a sperm or an egg divide by mitosis, that
is, by a doubling followed by one division, the daughter cell would
have the same number of chromosomes as the first cell, and when it
joined with its opposite gamete, the fertilized individual would have
twice as many chromosomes and die. Through meiosis, the cells divide twice, and one set of chromosomes, half the set of pairs, is an
appropriate mate for another half produced by another individual.
The cell is actually only half complete. Once the sperm meets the egg
and fertilization occurs, the new individual has two sets of the homologous chromosomes again, one full set from the sperm and once
full set from the egg. Genetic recombination has occurred as the two
sets of matching but genetically different chromosomes glide together.

   But why grab for the genes of others? Why use all this effort to
divide twice, lose half the pairs, and then have to search for a partner?
Why did sexual reproduction begin at all?

Why Have the Two Sexes?

   Various theories have been proposed as the force behind the giant
leap in nature to hard-wired sexual reproduction. Some authors have

suggested that sexual reproduction really doesn't confer an advantage at all, and that the search for meaning is futile. They believe that sexual organisms were favored for other, nonsexual reasons and that their mode of sexual reproduction came along as just so much baggage (Margulis and Sagan 1988, Williams 1975). This explanation seems intuitively incorrect; why would sexual reproduction evolve independently in four phyla and with such regularity if it conferred no advantage?

A group of molecular biologists see the evolution of sexual reproduction another way. They suggest that it is just a fix-it system (Bernstein, Hopf, and Michod 1988). Under this scheme, one set of DNA long ago took another set of DNA to repair DNA damage, much like the ancient bacteria when they swallowed their neighbors. Reproduction, or passing on genes, even mixing genetic material, was not important. The only patching that really counted was that which repaired the hole in the double helix of life. There's a small problem with this theory—one might expect cells to seek out other cells with closely matching DNA to facilitate the repair job. Perhaps they sometimes did. But such close matching also would have increased the bad effects of mutations carried when like steals DNA from like. The trade-off between mating with similar DNA and being wiped out because that similar creature carried the same bad genes along with the good ones probably makes this hypothesis less than reasonable.

The most common, and perhaps most accepted, explanation is that a varied pattern of genetic heritage helps an organism survive in an increasingly complex world (Weismann 1887). Natural selection operates by favoring certain individuals in a particular environment. When individuals reproduce sexually, their offspring are a patchwork of genetic material from each parent. Genetic recombination occurs when sperm and egg combine and the homologous chromosomes meet up. Each set of genetic material contributes to create a new individual with an entirely new genetic makeup. Natural selection, through differential survival and differential fertility, favors some over others. Without genetic variation, natural selection would have nothing to work with. It would have only a few kinds of clones, and once those were wiped out there would be nothing to select for or against. From the evolutionary perspective, genetic variability is necessary for selection to operate. From this point on, mutation, the process by which DNA itself changes, was not the only way to form new genetic combinations; the parents did it themselves.

What then does natural selection do with all this variation? Why might variation give sexual reproducers an important evolutionary advantage?

Life isn't static; organisms don't live in a world dedicated to their well-being. Take a species that's been honed to perfection by natural selection. Then make a drought one year, a flood the next. Change the plant life a little, introduce a new competitor. Mother Nature always sets a new stage in which selection operates, and the best organisms efficiently deal with changes of scene. Some researchers believe that genetic diversity gained through sexual reproduction gives parents a fighting chance to pass their genes on even if they can pass on only half their complement (Maynard Smith 1978). New genetic combinations might have unexpected advantages as genes from two individuals link and produce a surprise product. These new combinations have a shot at dealing with the unexpected. It's like a new roll of the dice with each batch of infants. This proposal assumes that some part of an environment will change more quickly than generation length and that clones will be at risk because they face situations the parents never had to deal with. Under this evolutionary scenario for the evolution of sexual reproduction, the costs of losing half of an individual's genes would be quickly outdistanced by the advantages of some flexibility in the offspring's genetic inheritance. Proponents of this idea often use a lottery example to illustrate their proposal. Let's say that there's a million-dollar lottery at stake. We don't know exactly what the winning number will be. One fool (our clone) buys seven tickets with exactly the same number on each ticket. Meanwhile, a more reasonable person (the sexually reproducing individual), one who unfortunately has less money (or fewer genes because of the cost of sexual reproduction), combines her money with a friend's and together they also buy seven tickets but with different numbers. Obviously, the second person has a better chance of making some money, even though she must share the winnings with her friend.

The problem with the changing-environment–varied-gene hypothesis for the maintenance of sexual reproduction is that it doesn't ring true in arenas where it should. Most asexual reproducers live in highly variable or marginal environments, and sexual reproducers often live in very stable environments (Glesener and Tilman 1978, Maynard Smith 1978, Whittenberger 1981). Sometimes sexual and asexual reproducers live in exactly the same environments. And then there's the cosmic question of what exactly a changing environment is, one full of winds and rain one minute and ice and snow the next? How drastic must the changes be to favor these genetically mixed-up sexually produced infants?

Changes in environment per se may not be the issue at all, suggest some biologists. It may be that the advantage to sexual reproduction is not a defense against environmental catastrophe, but a means of allowing offspring to explore and inhabit new territory (Maynard Smith 1971, 1978). As the argument goes, two populations move into a new environment. The organism formed by the joining of two parents, which now has a combined heritage, might do better because it's made up of equal parts of the exploring populations. Some offspring might get the best of both worlds and survive well in an area that's unfamiliar to both parents. The combination of genes might also be insurance against environmental changes. But this theory has problems too; how likely are organisms to explore new territory, and why would they? Would the advantages of being a hybrid in a new territory be enough to bring about selection for sexual reproduction or maintain it once it has begun?

The most recent addition to the genetic variability camp, one that's gaining favor, has been led by an evolutionary biologist, William Hamilton (1988, Hamilton, Axelrod, and Tanese 1990). Hamilton has suggested that we are overlooking a significant environmental factor that could quickly, within generations, select for genetic variability and thus sexual reproduction. Disease, he reasons, is a major selective force that might account for the evolution and maintenance of sexual reproduction. Viruses and parasites mutate much more quickly than other organisms do. They also infect higher organisms and are a major factor in death before their unfortunate victims reach reproductive age. Individuals who carry an array of genes inherited from two parents have, by definition, protection against a wide array of diseases and are presumably less susceptible than are clones with more basic immune systems. Thus a complex genetic background increases an individual's immune possibilities. The organism is like a country store stocked with small amounts of all sorts of goods. You might not need a button or a pickle today, but when you do need it, the goods are there. This theory also sets the stage for mate choice. If it's to your advantage to combine your genes with those of another disease-resistant individual, why not pay attention to clues that indicate excellent health? Given this scheme, sound evolutionary advice might include the directive—never marry someone with the flu.

Though the reasons are still undetermined why sexual reproduction evolved in the first place and why it has lasted over generations, it's still the favored mode for 99 percent of higher organisms. They may be resistant to disease or be able to exploit new areas or to survive environmental shifts. In any case, they produce offspring that aren't quite like either parent, infants with unique combinations of genetic material. As a result, most animals engage in the sexual act on a regular basis, and this fact causes us all sorts of problems. And so the human species spends endless hours thinking about sex and seeking partners right from the day we comprehend that there's such an activity available. Each sex will require different, often opposing, choices to make it through life. But these same differences make life interesting.

Vive la Difference

The first sexual act was a simple fusing of two equal-sized cells, sharing in the combination and division of chromosomes. Somewhere along the line, those original cells evolved into two distinct forms, those we call male and female. And then the system stopped at only those two flavors. We might have expected more. As G. A. Parker, an evolutionary biologist, asks, "Why not, say, five sexes, each producing its own characteristic gamete?" (1982, p. 281). But evolution so far has moved toward only two different gamete sizes and two corresponding genders. Natural selection, taking the minimalist approach, stopped at two sexes, enough to gain genetic variability for offspring without making the world more complex than it already is. Nature could also have opted for look-alike forms with similar gametes that join to create genetic options, but oddly enough the two genders most often look different from each other. How did this difference evolve, and why does Nature maintain this unequal system? Before we ask why females behave the way they do, we must ask why females are different from males in the first place.

Any human child knows the difference between males and females of our own species. Men are usually a bit bigger than women; they have facial hair and a penis. Women have breasts that stick out from their bodies and reproductive organs that allow them to gestate babies. But look at Canada geese, or even squirrel monkeys. You might not catch a glimpse of the penis, and you certainly wouldn't be able to see breasts. Most birds, except for waterfowl, don't have penes at all and only human females have such large pendulous breasts. Looking on the outside won't necessarily (or definitively) separate the two sexes for all creatures. How then would you determine which ones to call male and which ones to call female? You could take their blood and analyze its chromosomes to figure out which sex was which. Human females have two X chromosomes and males have an X and a Y, but this designation might not work for other groups. Male birds, for example, have two X's, whereas the females have the XY combination.

The only consistent way to separate males from females is to examine their reproductive products. Those creatures producing eggs are females, and those that produce sperm are males. The eggs are rather large, slow-moving cells, and they appear ripe and ready only at intervals. Sperm are tiny cells, although they contain a set of DNA similar to an the egg's, and their tiny size makes them the fastest cells on earth. And they're produced by the male form of a species at a relatively constant rate.

We can best explain the contrast between male products and female products by using humans as an example. Females are born with about 2 million egg cells, but fewer than 500 of them will ripen over a lifetime. Human females begin to make viable eggs at about age fifteen and stop at about age fifty. If all goes well, the human female ripens one egg every thirty days or so for a grand total of 470 fertile eggs. Take time out for pregnancy, nursing, and slowing down of the reproductive system during the last years, and a woman is left with only a few hundred chances of conceiving. This may still seem like a lot until we look at males. Human males also begin to produce viable reproductive products during adolescence, but their production line is a factory compared with the slow, careful construction of female eggs. Human males produce about 3,000 tiny gametes a day. The difference between eggs and sperm in size, shape, and production rate defines the sexes, and it has been around since the first eukaryotic cells merged in sexual play over 600 million years ago. The difference is easy to picture, especially when viewed under a light microscope. There they are, a large one with hundreds of small ones buzzing about. One form is round like a ball while the other looks like a tadpole with a large head and a wiggling tail.

Remember that the first sexual-like meeting of cells was a simple fusion between two equals. One encompassed the other, their DNA met, they divided and the predator used the DNA of the victim to repair itself. But very quickly, this balanced relationship gave way to exploitation of some cells by others. Once sexual reproduction was established billions of years ago and cells began fusing to create new life, tiny "male" sperm began to use large "female" eggs. The difference in the size of male and female gametes, called anisogamy to distinguish it from isogamy, in which gametes are of equal size, is now an established fact of evolution, and the historical reason for all our sexual problems.

The fact that male sperm are smaller than female eggs in every animal is so common that we barely notice this fundamental difference defining males and females. But the disparity in gamete size, shape, mobility, and production rate acts like the top tile in a pile of dominoes. The first domino to fall was the "sexual" act itself, when one cell merged with another. Then at some point one cell exploited the other, and two separate stands of dominoes began to fall in opposite directions. Most scientists agree that the initial exploitation was a matter of nutritional need. Some small cells responded to the opportunity of nearby larger but similar cells. The smaller, mobile ones could swiftly reach slower large cells, break into them, and utilize their nutrition and DNA. The fast-track small gametes divided quickly and turned into millions of squirming cells propelled toward their slower fat victims. This strategy was so successful that natural selection quickly favored the production of those tiny cells. And why did the large cells allow this exploitation? At first glance, they were not doing as well, but keep in mind that the large cells also passed on genetic material. These large cells also had a viable means of passing on genes. They never wasted energy running around as the tiny gametes did. Instead, they invested all their energy into producing fewer copies of nutritious cells (Whittenberger 1981). They floated about, retaining their nutrition, not wasting a movement, available for any mobile gamete. But as one biologist, Ute van den Bergh, puts it (personal comm.), each type of cell utilized a good strategy, and only after it was too late did each evolve to polar opposites that were forever dependent on each other for reproductive success.

How exactly all this size differentiation started, where the big and little cells came from in the first place, and how they turned into one form or the other, no one really knows. Maybe some medium cells became smaller and others became larger. Or perhaps they were all the size of eggs and some were selected to become much smaller; or, vice versa, all were originally the size of sperm but someone had to carry the nutritional end of the deal. We don't know whether the original isogamete was small or large and the opposite form selected. We only know that selection favored the difference. In any case, one form became small and fast with no ability to sustain itself nutritionally. It therefore needed the other form, the slow-moving, stable, nutritionally fat egg. Keep in mind that both forms had an equal number of chromosomes and that their relationship, born of nutritional exploitation of the larger by the smaller, was, in the end, cooperative. If the big fat egg allowed the skinny mobile sperm to join it, the recombined cells divided into an offspring. Most likely sperm were responsible for the initiation of this joining because they needed the egg's food supply, and the more they needed, the faster and more streamlined they became, competing with one another for the bulbous eggs (Margulis and Sagan 1988).

Recently, evolutionary biologists have suggested another possibility: the unequal-sized sperm and egg system is really the result of the surrender by one type of cell to another as a means of avoiding intracellular conflict (Anderson 1992, Hurst and Hamilton 1992). They suggest that when the two ancient cells joined up the chromosomes readily lined up to create a new individual. But other features of the two cells, their organelles, would be at war over the only available space in the cytoplasm. Think of it this way—a couple moves in together, happy to be sharing a bed, but together they have enough furniture for two households. Something has to go. Hurst and Hamilton believe that one original cell opted to jettison its furniture, in this case its mitochondria. Thus we have small sperm without mitochondria and larger eggs that are responsible for all the mitochondria in sexually reproducing species.

In any case, this unequal process with two different forms of reproductive gametes is such a successful cooperative/exploitive venture that equal-sized sexually reproducing gametes are found only in a few fungi and protocists. Almost every form of sexually reproducing animal and plant life on earth today lives in a world utilizing two, and only two, sexes with different-sized gametes.

We also know that the organs needed to produce those two kinds of gametes are, in form and design, different. Males have testes that manufacture sperm at a continuous and rapid rate. Females have ovaries that slowly mature eggs and release them at intervals. More difficult to comprehend than this basic biology is the profound effect of that difference on everything that males and females do. One small nudge billions of years ago toward an inequality in gamete size, and everything about maleness and femaleness clicks over until we have two distinct genders, two forms of one species distinguishable not only by their gametes but also by their looks and, more startling, distinguishable even in how they behave.

The differences in form and size between the two sexes, beyond gametic differences, is called sexual dimorphism—two morphs or forms based on sexual differentiation. One sex is usually larger in whole body size than the other (most often males). In some species, the sexes come in distinct colors too. Some birds, for example, are so differently colored that males and females have been mistakenly classified as two different species. Sometimes special characters appear in one sex and not the other; sometimes one sex has the more elaborate version of a similar trait. Male peacocks wave their extravagant plumage while female peahens look rather drab. Male savannah baboons have canines that could rip the flesh off a horse, and female baboons in comparison have useful but not very ferocious teeth.

For some, the notion that males and females act differently is the most difficult biological lesson to swallow. Many of us would like to believe that there are no behavioral differences between men and women. Certainly in patterns of intellect and mental ability, few striking genetic differences between the sexes have been documented. But there are other important distinctions between males and females in general, ones that shouldn't be particularly offensive in their proper evolutionary context. For example, among primates, males are frequently more aggressive, showy, ready to threaten and chase than are females. Females are sometimes more social than males. Anyone who's watched monkeys in a zoo would see this difference among the adult animals. Yet females are often competitive and aggressive when need be, and males are interested in social interactions too. In other words, although males and female show differences, there's much overlap in their behavior, but these differences don't point to male or female monkeys as smarter or more clever than the other gender. They are just driven by different strategies to get along in their monkey world. Males and females have different life-history strategies, different ways to achieve survival and reproduction.

The point is, because of basic biological differences at the gamete level, males and females have different patterns of behavior. Males have those expendable gametes and the potential to spread their sperm about. Females have only so many chances to pass on genetic material in the form of infants. They must be more careful and guarded. From day one, their investment tallies up more than does males'. One way is not better than the other; one is not more efficient, more graceful, or even necessarily more successful. The ways are just different. It doesn't matter if you are a toucan, a lion, or a gorilla, your gender dictates how you go about life. And because males and females have different reproductive interests at heart, conflict between the sexes is the rule, and compromise a must.

Female Primates in Particular

From eggs and sperm, the evolutionary story moves on to individuals. At that level, males and females engage in conflict and compromise to gain reproductive success. Within the context of the battle of the sexes, females make their reproductive choices. Female mammals in particular are bound by the constraint of making mature nutritional eggs and the stark reality of intense parental investment, months and years of lactation and care. The game of reproductive success becomes much more complex because of a style of reproduction that includes lactation. Lactation pushes even further the polar differences required by males and females to pass on genes—it places the burden of parental investment heavily on females. These high-intensity female mammals do, however, have options. Within a life pattern typical for a species, each female makes conscious and unconscious choices. She might breed early, have few offspring, and die young. Or she might have huge litters and live the typical life span. Her path is constrained by what physical body she was born with, how that body interacts with the environment, and what it takes to pass on genes.

Within this context, primate females are more than interesting subjects. They're at the deep end of maternal investment; all primates mature late compared with most other mammals, they have long life spans but reproduce only at intervals, no more than two infants appear at a time, and primate females invest heavily in each one. In this book, I use nonhuman primate females to illustrate principles of female sexual behavior in two ways. First, this material is presented to provide comparisons: we see that female humans are not as special as we might imagine. Many of the mating strategies described in the chapters on nonhuman primates will be familiar to the men and women reading this book. The second purpose is to tie all female primates together into one large sisterhood. That concept of sisterhood is crucial to our understanding of the nature of human female sexuality and mate choice.

We must watch other primates and learn from their behavior about our roots. We can understand what we're made of, and how we operate, only if we look at all the available information, and part of that information lies close at hand in our nonhuman primate relatives. Put more simply, our place in Nature is defined only by how we fit into the larger picture of primate nature.

CHAPTER 2

Primate Cousins: Looking into the Mirror

MOST PEOPLE AGREE that humans are by far the most fascinating of animals. Humans construct complex civilizations, mold their lives around myths and religious beliefs, and do bizarre things like wear clothing, paint their bodies, and sing. These cultural accoutrements, some might add, are the most interesting facet of our behavior. Indeed, culture shapes many of our likes and dislikes, and we learn from other humans about socially acceptable behaviors and traditions that help us get along in human groups. But has culture really broken the chains of DNA so that it alone is now responsible for human behavior? Are humans so different from other creatures that they receive a special dispensation from their biological heritage? And most important, is our cultural fluff really the most fascinating aspect of our behavior?

I agree that culture certainly colors how we act and who we are. But it doesn't take away from the universal nature all members of our species share. Humans are basically no different from other animals. Just like all other species, we eat, move about, reproduce, and die. Like all other animals, we also have an evolutionary past, a rich biological history that informs the present and will affect the future. The one thing that separates us from other animals is self-consciousness: so far the ability to question existence seems to be a uniquely human characteristic. But it's also a burden. As self-aware creatures, we have no choice but to look into the past and attempt to figure out where we came from.

Years of research on nonhuman primates suggest that there are strong behavioral connections between ourselves and the other members of the primate order, including prosimians, monkeys, and apes. Our humanness isn't based solely on what culture feeds us; it also contains a biological essence that influences our behavior. Natural selection has acted on our physical appearance and our behavior in many of the same ways as it has on other groups of animals, especially other primates.

The road from our past into the present weaves and bends around physical characteristics that now define us as human. Our species is typified, even defined, by large brain size, bipedalism, the use of tools, and language. But before those specifically human traits evolved, we were just basic primates, apes about to embark on a divergent evolutionary path. We still have many of the same original primate traits, but like other species that branched off, we added a few more that differentiate us from our primate cousins.

Our ancient ancestor species did not look like present-day chimpanzees and it did not look quite human; it probably had many characteristics seen in both species today as well as traits eventually selected against and eliminated in both species. For some reason, two lines, one chimpanzee and one human, were selected and the two types parted ways. Chimpanzees evolved along their own route, while more humanlike creatures eventually stood up on two feet and walked across the savannah, beginning the singular journey of human evolution.

We know that the physical appearance of modern humans evolved over millions of years. Paleontologists have been able to trace the human lineage back four million years by using a fossil record of skulls and skeletal parts. They've tracked the evolution of our big brains and changes in the shape of our faces. But the paleontologists can't trace the evolution of human behavior which also evolved over millions of years with the rest of our bodies. Bones can't tell us how protohumans interacted or, more important, *why* certain patterns of behavior were selected over others. For the path of behavioral evolution, we have to use more indirect evidence. Today's primates, and their various behaviors, provide a living museum of our past. Prosimians, monkeys, and apes are the "fossil evidence" that provides clues about the behavioral evolution of the human species.

The Primate Path

Our evolutionary tree is inhabited by such creatures as lemurs, lorises, baboons, and gorillas. It shakes and bends as species have been, and still are, modified and molded by natural selection. In a broad sense, prosimians, monkeys, and apes, the other members of our primate order, are examples of what we once were. Certainly they've evolved down paths of their own design, but this track doesn't take away from their role as our forebears. As monkeys moved away from prosimians, as apes moved away from monkeys, and as humans moved away from apes, we all took along copies of the basic genetic material that we still have in common today, and this commonality is reflected in shared features of our anatomy and behavior.

This living genealogical tree is different from the evolutionary history of most animal orders. For example, if we had turned out as horses with an interest in discovering ourselves, the behavioral past would be lost to us. In the long history of horse evolution, many species of horses died out, leaving truncated lines of evolution. Luckily, we humans come from an order that even today contains living representatives of the various stages of our primate past. The route of human evolution, racing forward through time, passed though a prosimian phase, a monkey phase, and an ape phase, but unlike other animal lineages, each primate group didn't fade into oblivion (Fleagle 1988). For example, monkeys evolved from prosimian forms about 25 million years ago during the Oligocene epoch. From these ancient monkeys, some sort of ape evolved, and eventually humans appeared. But the monkeys of the Oligocene didn't die out; they became the monkeys of today and preserved that ancient stage of monkeyness intact. Instead of becoming extinct, our primate ancestors maintained a biological integrity all their own. They branched off and clung tightly to the evolutionary tree. As a result, today the forest and savannah are populated with contemporary examples of our evolutionary history, living fossils that screech, scream, and bite. Their behavior isn't exactly like that of ancestral primates living millions of years ago, but today's monkeys do provide precious clues about our primateness. Living primates are our windows to the past. Because our phylogenetic relatives are close at hand, we can lift the sash, let in the cool breeze of evolutionary history, and look around for the essence of the human character. It's nestled within a generalized primate nature all around us.

It is reasonable for scientists to ask primates to dress up as our bygone relatives and take part in our human evolution play because of a straightforward biological fact—we share genes with today's other primates. The relationship between other primates and humans isn't just speculation on the part of eager anthropologists. New methods in genetics allow for direct comparison of the tiny strands that makes us what we are—deoxyribonucleic acid, or DNA. Certainly primate genes have been shuffled and reshuffled, and some have been lost in the game of reproductive success. But all primates share some genetic material in common. Geneticists now estimate, for example, that humans share 98 percent of their genetic material with chimpanzees (Goodman 1963, Sarich and Wilson 1967). Unsettling as this may be to some, most of us realize the truth of this connection when we gaze at chimpanzees in zoos. How can we dismiss the similarity in body build and the way they look at us as if they "know"? Like a three-stage rocket launched into space, the path of human evolution ejected genera and species along the way to a human orbit, and those creatures developed along their own trajectories. But the rocket left a trail of DNA which lights up the evolutionary sky. The human genetic alliance with other primates is a passport to our heritage, and prosimians, monkeys, and apes are the visas that allow an investigation into the history of human nature. Because this connection to our primate cousins is so strong, we can use these animals to understand how we evolved into humans.

In this book, I specifically use primates to ask questions about our sexual forbears. It's not enough to suggest that human females may have particular mating strategies. We must know how common those strategies are and, put into a larger primate context, what the patterns mean in an evolutionary sense. This book eventually looks at female humans as it looks at other primates. In my view, human females are just another kind of ape with sexual and reproductive interests in mind.

In Search of Primitive Primates

Nonhuman primates today are restricted to a belt of geography which closely adheres to the equator. It's incorrect to say that all primates are tropical-forest animals, although the media usually depicts monkeys and apes that swing through jungles. To be sure, the presence of trees is almost always a necessity for primates because trees are the best escape route for most. But many species live in temperate zones, some travel over rocky cliffs, while others battle the cold. For example, Japanese macaques, famous for their lazy afternoons in hot springs, spend much of the winter high in the snowy mountains. Other primates, including baboons, live in large groups roaming the African grasslands under hot sun without the closed canopy of a forest. Gorillas rest quietly in bamboo forests on the slopes of volcanic mountains, and human visitors are surprised at the cold, wet climate and the mist that envelopes gorilla groups. Woodlands too are often home to monkeys who flip from branch to branch or scurry through the undergrowth. Sometimes silent, sometimes noisy, groups of primates inhabit the forests of Africa, Asia, and Central and South America.

Prosimians, or strepsirhines, the most primitive of primates, live only in Africa and Southeast Asia. They can be most easily divided into three groups, lemurs, lorises, and galagos. They are distinguished from more advanced primates by some physical features that highlight their basic mammal status (Bearder 1987). A strepsirhine primate has a wet nose, called a rhinarium, just like that of a dog or cat. The rhinarium signals that the sense of smell is still important to these creatures, and perhaps that the sense of sight has not been so highly selected. Head on, a strepsirhine can also be distinguished by the split in the upper lip. As with a dog, the two sides of the lip meet in a curved upper sweep just under the wet nose. Strepsirhines also have independently movable ears, which help locate insects for food. Although primates always have nails on most of their fingers, sure signs of primateness, strepsirhines still have some claws. In particular, they retain one claw used for grooming, like a toothpick glued to a hand or foot. A strepsirhine passes the elongated claw though the fur of a friend or contorts its own body to scratch an almost unreachable place. As an aid in grooming, these ancient relatives also have a set of bottom front teeth that form a comblike structure. The lower incisors are long and thin and closely aligned. With this "tooth comb" they brush one another or themselves.

Many strepsirhines are active only at night, and thus they are difficult to find and even more difficult to study. The best way to locate them is to grab a flashlight and creep through the dense undergrowth of Africa or Asia—only hardy souls need apply for this job. You don't use the flashlight to guide your way: your feet are on their own. Aim the light into the bushes and sweep it slowly up and down, back and forth. Suddenly, two tiny red Christmas lights appear in your beam. Success! You may have stumbled on a nocturnal strepsirhine primate, and if it stays frozen in your light, you get a chance to check out its size and shape. The red dots are the eyes, and they show up in directed artificial light just as the eyes of deer light up on dark highways. The retina of the eye has an extra layer that reflects light and recycles it through the field of vision. Like a flash on a camera, this recycled light makes it easier for the animal to see at night. All strepsirhines, but not monkeys, apes or humans, have this primitive eye with a reflecting tapetum, regardless of their daily habits.

Lemurs, one large group of strepsirhines, are found only on the island of Madagascar. About 120 million years ago, when Madagascar split from the African continent, strepsirhines ruled the land. Monkeys hadn't yet evolved, and apes were far in the future. Eventually, monkeys and apes would take over most of the possible primate niches in Africa, but not on Madagascar. There these primitive pre-monkey animals became ultimate opportunists in their new island home. Because the evolutionary path of this primate was geographically separated from that of all other strepsirhines in Africa, it didn't experience the same selective pressures. As a result, Malagasy lemurs have stayed basically as they were when the island split off, with of course a few modifications over the millions of years, and a few species, at least fourteen, also became extinct (Richard 1987). Without monkey competitors, the successful lemurs radiated into every ecological niche. They took over forests and spread into the dry deserts. They leaped out of the trees and took over the land. But their brains never grew to monkey size, nor did they develop the short monkey face or a quadrupedal way of walking. Today they look essentially as they did millions of years ago, dinosaurs of the primate world.

Lemur species now come in many shapes and sizes, including larger-bodied ringtails and brown lemurs as well as the miniscule mouse lemur. They come in all sorts of colors too, brown, white, black with males and females sometimes wearing different coats, often confusing even the best taxonomists. Lemurs also command the forests of Madagascar with wild cries and uninhibited jumping behavior. Malagasy primates move about the forest as if they were mounted on pogo sticks. They begin flight in an upright position; a lemur clings with its flexible primate fingers to the side of a tree, looks about, and judges the distance. Then, compelled by the need to move on, it pushes off with powerful back legs and leaps through space almost vertically, arms spread wide to keep balance. It lands on target, often in an upright position, and pokes its head around the tree, calculating the next tree-to-tree gap. The only thing missing in this locomotor pattern is the boing-boing-boing sound that should accompany its pogo-stick propulsion.

As strepsirhines radiated out over Madagascar, they adapted both physically and socially to new environments, and that radiation is reflected in their diverse ways of living. Group size varies: some species are rather solitary nocturnal animals, others are daytime species living in communities. The nocturnal lemurs, like the lepilemur and the mouse lemur, travel alone at night, meeting up only to mate. Diurnal, day-living lemurs may be solitary or live in small family groups or in large troops. The social lemuriforms also exhibit a system not often found in many mammals—females are dominant to males (Jolly 1966, Richard 1987). Females have first priority at feeding sites and clearly dominate males in social interactions.

The other strepsirhines, galagos and lorises, are native to Africa and Asia (Bearder 1987). Galagos, with large red eyes and ears like radar screens, are the nocturnal squirrels of Africa. They live mostly on insects and use sound to locate and capture prey. Mothers park their babies in nests and move about the dark bushes, rustling up an insect dinner. Lorises are more circumspect; their lives run in slow motion. In the tropical jungles of Asia, they lazily sneak up on large insect prey, hopeful that the bug won't move away first. We know little about most of these particular nocturnal strepsirhines because they are so difficult to observe. To shine a flashlight in a loris's face is like paralyzing it in a searchlight—no nocturnal primate readily carries on with its business.

Although we know less about the strepsirhine primates than about the more closely related monkeys and apes, they do provide the first keys to understanding why we evolved along a certain direction. We see in strepsirhines a brain that may be smaller than that of monkeys to come, but larger than the brains of all other mammals, and many of the lemurs have sophisticated social lives that resemble monkey societies in complexity (Jolly 1966). They aren't really "primitive"—

this is just a judgment call from our human perspective. Strepsirhines are just more distantly related to us than are the more familiar monkeys and apes. But they poke their pointed faces through our evolutionary curtain to remind us that even though we are on our way to large-brained specializations basically we are still mammals.

The First Transition

It doesn't look much like a missing link. It has a small body, huge eyes, and feet that would shame a kangaroo. It's not particularly social and it eats bugs. Regardless of looks, the tarsier, a strange creature from Asia, may be a connecting bridge between strepsirhines and more specialized monkeys and apes. The special position of tarsiers is confirmed by their taxonomic position: tarsiers are placed not with strepsirhines but with monkeys and apes into a group collectively known as haplorhines.

A clue to the tarsier's special status among primates is its brain size. The tarsier's brain is a bit larger than a lemur's, but not quite monkey enough. Also, the tarsier exemplifies natural selection's push for good vision in primates; it has huge eyes and can see well. John Fleagle, a primate anatomist, calculated the size ratio of tarsier eyes and brains and discovered that each eye is actually larger than its brain (1988), a ratio suggesting that vision for this creature is more important than smarts for keeping alive. We don't know much about the behavior of tarsiers. Some, like the Bornean tarsier, are rather solitary, and some, like the spectral tarsier, live in closer contact and forage mostly in pairs (Bearder 1987). But taxonomically the tarsier is more like a monkey than like a lemur. It has no reflecting eye, although it has good night vision, and the tarsier sports a dry nose and continuous upper lip like a monkey's. This tiny creature is a steppingstone along our evolutionary path because the tarsier shows the beginnings of more monkeylike adaptations, especially a trend toward increasing brain size. From this illusive Asian creature, we move directly to animals that share so many traits that they are direct mirrors of ourselves—monkeys and apes.

A Barrel of Monkeys

Most people think that all primates are monkeys. But there's nothing more annoying for a primatologist than listening to the crowd in front of the chimpanzee cage at the local zoo and hearing, "Look at the monkeys." I find myself correcting perfect strangers with, "They're not monkeys, they're apes." This misclassification annoys those in the know because of the meaningful distinction between monkeys and apes. This distinction is also important to the story of human evolution: we are more closely related to apes than to monkeys. At the same time, much of what monkeys do pertains to ancient patterns of human behavior we still share with these more distant cousins.

There are two kinds of monkeys, those from the Americas and those from Africa and Asia, or the Old World. American monkeys, also known as New World monkeys, are an anomalous lot found in Central and South America. No one knows where they came from, why they're in the Americas, or how they evolved down paths so different from those of their close phylogenetic relatives, Old World monkeys.

Long before full-fledged monkeys roamed the southern parts of the Americas, there were ancient lemurs in North America. But these prosimians died out and never evolved into monkeylike animals (Fleagle 1988). As a result, there's a long gap between those ancient lemurs and the monkeys who appeared about 30 million years ago in South America. Where did these New World monkeys come from if they didn't evolve directly from the North American strepsirhines? The monkeys we now see in South and Central America probably came from a common monkey stock that drifted from the African continent over 30 million years ago when the two continents were closer together and the water level of the South Atlantic was much lower. Although this view calls up visions of monkey armadas looking for a better land, probably a simpler ecological situation facilitated their move. Paleontologists speculate that large pieces of marsh broke off from Africa and drifted west. Monkey populations who happened to be living on those land masses then colonized their new land and began a process of parallel evolution which mimicked, in many ways, the concurrent evolution of their Old World relatives.

There were no strepsirhines in the new land, and no apes ever evolved in the Americas either. So today the monkeys occupy many niches filled by the apes or strepsirhines of Africa and Asia. Today monkeys can be found in tropical jungles of Costa Rica, Belize, Honduras, and Panama in Central America and in Brazil, Venezuela, and

Peru in South America. The Amazon Basin and a small patch of forest left on the eastern Brazilian coast are the last stands for several highly endangered species. New World monkeys are often difficult to find, because all of the species are arboreal; that is, they live high in the treetops. In addition, they've been regularly hunted by indigenous peoples and are generally wary of humans.

You can tell a New World monkey from an Old World monkey by its nose. New World monkeys have widely spaced nostrils with a large area of skin in between, and the nasal openings are directed to each side (Hershkovitz 1977). Old World monkeys, on the other hand, have narrow noses and the holes are oriented downward. Some American monkeys also have something precious that Old World monkeys lack—a prehensile tail. The tail is made up of muscles and is used like a fifth limb. American monkeys with a prehensile tail, such as the capuchin or organ grinder's monkey, hang from trees by their tails or they use their tails like a hook to pull branches. And because all New World monkeys are arboreal, those with prehensile tails are all the more lucky because their tails are useful in arboreal locomotion. As good primates, American monkeys have nails on their highly flexible hands. But marmosets and tamarins, tiny brightly colored monkeys, still retain some claws that aid them in scratching through trees like squirrels.

Some species of New World monkeys, specifically marmosets, tamarins, night monkeys, and titi monkeys, live in small family groups and breed monogamously (Wilson Goldizen 1987). Monogamy is very rare among mammals, but there are more pair-bonded species among primates than among any other order of mammals. Most of these monogamous primates are New World monkeys. There's no way to know what part monogamy played in the evolution of human mating systems (I offer some speculations later), but New World monkeys are invaluable to our quest for understanding a pair-bonded intimate mating system. These small groups of breeding pairs and offspring maintain small territories they defend with high-pitched morning calls and vigorous fighting when they encounter other small groups. Monogamous species also include the only nocturnal monkey, called the night monkey or owl monkey (Wright 1978). Some medium-sized monkeys, such as the capuchins or sakis, live in moderate-sized groups of ten or so animals. Small-bodied squirrel monkeys travel in large groups of one hundred animals and announce their presence with arboreal acrobatics and noisy crashing through the canopy on their way to a fruiting tree. Howler monkeys, who come in black or red varieties, fill the forest with howling cries and have been called the loudest creatures on earth (Crocket 1985, Glander 1984). Every morning, large groups of howlers howl to announce their presence in the forest. They yell, "This area is ours, stay away." Unlike the gaily colored marmosets and tamarins, or the raucous howlers, spider and woolly monkeys maintain a dignified presence in the forest, far from the curious eyes of most observers. They live in small groups at the highest reaches of the canopy (Robinson and Janson 1987). They are large bodied, like small apes, and display great dexterity in moving about the forest by swinging under branches rather than leaping from branch to branch.

But it's the colorful palate and odd physical accessories of American monkeys which catch the eye. Golden lion tamarins are really bright gold; cotton-topped tamarins sport white Mohawks that give them a punk flare; and squirrel monkeys wear Zorro masks on their faces—they look like bandits. The dense jungles of the Amazon rustle and shake with movement—everywhere is the scurry, leap, and crash of brightly colored pint-sized monkeys squeaking at strangers.

Monkeys of the Old World

Although American monkeys are related to humans, more so than any strepsirhine, the path of human evolution must leave them behind and reboard the monkey raft and return to the Old World. Across the oceans from South America, in both directions, are scores of Old World monkeys. They may not be as closely related to humans as are apes, but Old World monkeys are important players in our historical quest. About 25 million years ago, all monkeys and apes shared a common ancestor. Although apes, and later humans, evolved from this generalized monkey stock, the Old World monkeys continued their successful radiation across the globe and retained monkey ways. Today Old World monkeys include over 70 species ranging across Africa, including North Africa, through Asia and southeast Asia. They all have those downward nostrils and none has a prehensile tail. We share some physical features in common with all Old World monkeys—for example the same number of teeth—but our real connection is in the social realm. We recognize ourselves more in what they do than in how they look.

Monkeys of the forests and savannahs in Africa and Asia display a remarkable variety of group compositions, from small family groups hidden in tropical forests to monkey battalions roaming the African savannahs. The most commonly known—macaques, vervets, and baboons—are probably the most numerous. We also know more about the social habits of these three species than about any other monkeys. The reasons are simple: they live in large social groups and are easy to watch. They're also the most fun to observe—something's always going on. What looks like a field of medium-sized brown monkeys peacefully munching on grass is really a seething story of interpersonal intrigue. Every monkey knows who's who in the group. And every monkey keeps score of personal favors and attachments, and the whole group watches one another minute by minute. Macaques, vervets, and baboons live lives that shame any miniseries.

Baboons are found only in Africa but they inhabit all kinds of ecological niches, including dry savannah, woodlands, and the harsh steppe of Ethiopia (Altmann 1980, Altmann and Altmann 1970, Dunbar 1984, Kummer 1968). Macaques are world travelers. Long ago macaques left Africa, leaving behind only the Barbary macaque in North Africa, and moved into Asia. Now macaques inhabit forests and woodlands in India, Pakistan, southeast Asia, Indonesia, Japan, and the Philippines (Lindburg 1980, Roonwal and Mohnot 1977). Macaques, with 15 species and a widespread distribution, have successfully survived human encroachment and habitat destruction. Vervets live in the eastern part of Africa and have been successful in national parks. Dorothy Cheney and Robert Seyfarth, primatologists, have used sophisticated recording techniques to test what vervets "know" about their social environment (1990). Vervets, and presumably all Old World monkeys, have a sophisticated system for keeping track of relatives, enemies, and friends, and they are much better at knowing their social world than they are at paying attention to their environment.

Less well known are the African forest Old World monkeys, such as guenons, mangabeys, blue monkeys, and redtails (Cords 1987). Some live in one-male groups from which a resident male is often ousted by an invader. And these monkey groups often live in overlapping patches of forest, stratified in the canopy and on the ground, making sure to rely on different resources to avoid feeding competition.

The Generic Monkeys

Perhaps the best-known primate is the rhesus monkey. It has not received the media coverage that chimpanzees have, and certainly it's not the most attractive primate around, but rhesus monkeys have served to illustrate much of what we know about primate nature. Primatologists agree that right on down the phylogenetic tree all primates have something to say about our heritage, and rhesus, and other macaques, have been instrumental in teaching us what makes a primate a primate. Because of their ability to survive well in captivity, rhesus are used extensively in biomedical research. All over this country, thousands of rhesus currently live in single cages, small enclosed houses, and large field cages. For generations, they have served as human models of our past and guinea pigs of our present. It's no wonder that some of the best information on female mating patterns, and female choice in particular, comes from this unassuming monkey.

Rhesus monkeys are found in great numbers in India, Pakistan, Thailand, and other parts of south Asia. Tourists know them best as the irritating monkeys at temples which steal food and act as though they would just as soon bite you as groom you. This brownish medium-sized monkey has short hair, long pointed canines, an upside-down triangular face, and a rather short tail. Just by chance, the rhesus is fixed in the average person's mind as what a monkey looks like and acts like—the generic monkey.

Field studies by Indian and American primatologists have been important to researchers in the United States because of the use of rhesus and other macaques in laboratory research (Lindburg 1971, Southwick, Beg, Siddiqi 1965). For decades, rhesus monkeys have been used for the growth of polio vaccine and have acted as subjects in drug testing and in research on infant development. Rhesus also serve as subjects of behavioral research by anthropologists, psychologists, and zoologists. And it's these studies that have given us long-term data on how rhesus life histories unfold—and how those lives relate to human behavior. The rhesus system involves large groups constructed around powerful matrilines, male emigration at puberty, intense social interaction among individuals, seasonal breeding, and the absence of male care of infants. Macaques in general are the cockroaches of the primate world—and I mean that in a positive way. They can adapt to anything—changes in weather, high altitude, and, most important, human encroachment. Rhesus in particular have been very successful adapting to the ways of humans and thus they are not endangered, and probably never will be.

Although many behavioral studies outside of India were, and still are, conducted in laboratory settings, perhaps the most interesting information on rhesus social behavior comes from almost fifty years of observation on a unique population. In 1938, Clarence Ray Carpenter, a zoologist and animal behaviorist, had an idea. He wanted to form a troop of rhesus monkeys within easy access of the United States, a population that could be followed for rhesus generations and used only for behavioral research. Carpenter captured monkeys from India, and three months later 409 monkeys were dropped on a small uninhabited island, called Cayo Santiago, off the coast of Puerto Rico (Rawlins and Kessler 1986). The animals could roam freely on the island's thirty-seven acres. Although admittedly not much like the temples, forests, and villages of India, Cayo's forested land offered protection from humans and predators. Since 1940 this population has been watched by scientists, census takers, and students. Every kind of noninterventive research has been conducted on these animals, from studies of mother-infant behavior to studies of emigration (Rawlins and Kessler 1986). Some researchers know these animals as well as they know their own families—probably better. They have recorded genealogies and histories of how rank is established and changed over time, and now paternity is being identified with DNA markers (Berard et al. ms.). There are trade-offs, of course. The animals are fed, and water is provided, thus asking questions about rhesus ecology and feeding strategies is inappropriate. There aren't any predators on the island and no monkeys can migrate to another part of the country—there's no escape. At times the population count is higher than any rhesus in India could imagine. Surely this abundance changes the quality and quantity of social interactions, by the mere fact that each monkey has so many possible partners. But Cayo Santiago still provides some of the best data on rhesus, or any monkey, social behavior.

Another kind of macaque has also provided extensive information on macaque social systems, the Japanese macaque (Huffman 1991a).

Japanese macaques are the burly members of the genus; they are large bodied and covered with thick gray fur, which makes them look even larger. Adults have bright-red faces, and during the five-month mating season the red faces of females turn even brighter. In the 1950s, three sites in Japan were organized for behavioral research, Toimisaki, Koshima, and Takasakiyama. Although the macaques had previously been hunted and were wary of humans, researchers gained the monkeys' confidence by provisioning. Some groups live high in the Japanese hills, while others have been coaxed down onto beaches. Its from these studies, especially the long-term work on Koshima Island, Takasakiyama, and at Arashyiama, that we've gained a detailed account of macaque social structure (Imanishi and Altmann 1965, Itani 1959, Kawai 1958, Kawamura 1958, Takahata 1982a). Like rhesus and other macaques, Japanese macaque social structure is formed around matrilines, lines of descent through females. Some matrilines are dominant over others, and high rank gains feeding and reproductive privileges. When groups fission, they do so along kin lines. Young males emigrate at puberty while females stay with their mothers and female relatives. The monkeys form a tightly knit social group, and behavioral interaction is the most important feature of their life-history pattern. Japanese macaques are seasonal breeders, and females produce one infant each year or every other year. Today researchers are monitoring both provisioned and wild groups, and the data continue to pour in. The details of the male and female mating strategies of this species have helped explain how mate choices influence more general aspects of social structure. In that sense, Japanese macaques and their close relatives, the rhesus, have shown us that mating can't be disconnected from general social interaction; the way animals go about making new macaques influences how males and females interact even outside of the mating moment.

The Leaf Eaters

There's a class of Old World monkeys less well known than macaques and baboons who have adapted to a completely different ecological circumstance—the colobines. Colobines, which include the black-and-white colobus, langurs, and the long-nosed proboscis monkey of Kalimantan on Borneo, have the special ability to digest large amounts of leaves. Their molar teeth are especially pointed; like the blades of a food processor they pulverize leaves, thus helping to break up the fiberous structure of leafy matter and begin the extraction process needed to absorb such nutrients as protein. Then their specially designed chambered stomachs make for slow passage of a leaf mass. Leaves aren't just difficult to digest, they are full of poisons called secondary compounds. Colobines in particular are able to expel most of these poisons and absorb only good nutrients. Among the leaf eaters are Hanuman langurs, the sacred monkey of India. In Hindu mythology the original Hanuman was a monkey who served the god-king Rama as a loyal servant. Today this less-than-godly silver monkey raids crops and invades garbage pails, but in appreciation of Hanuman's past loyalty, the Indian people treat the pesky monkeys with benign tolerance. Langurs have been especially enlightening to primatologists because of one dramatic behavior—infanticide (Blaffer Hrdy 1977, Hausfater and Blaffer Hrdy 1984). When a male langur invades a one-male troop and takes over from the previous resident, he often kills infants. This seemingly odd behavior is actually a male reproductive strategy because it quickly readies a female langur for mating by bringing on her period of heat. Langurs were the first to demonstrate that infanticide, a seeming abhorrent behavior, can be an evolutionarily advantageous strategy.

In general, Old World monkeys are most important for research because of their diversity, rather than because of their commonalties. The array of social systems, mating styles, and ecological adaptations of Old World monkeys shows that evolution doesn't follow a straight and narrow path. Each niche, each change in habitat, molds species and brings forth differences. Old World monkeys are a celebration of Nature's variety.

The Ape without Us

The last group of primates, the apes, is probably the most familiar. Who hasn't followed Jane Goodall as she continues her exceptional life among wild chimpanzees? And who missed Sigourney Weaver's performance in *Gorillas in the Mist* as Dian Fossey, gorilla woman of Rwanda? The apes are our closest relatives and thus they have a special place in our hearts.

There are two kinds of apes, the "lesser" apes which include gibbons and siamangs, and the "great apes"—the orangutans, gorillas, and chimpanzees. They all have some physical features in common. For example, the best way to tell an ape from a monkey is from the back. Apes don't have tails. If you can get a backside view, sophisticated taxonomy becomes child's play. Apes also have body parts and ways of moving, that distinguish them from monkeys. Remember the jungle gym at the local park, the so-called monkey bars? They should be called "ape bars." If a decent monkey were to leap on the bars, it would immediately run across the top like a quick-footed cat or dog. Monkeys, like cats and dogs, are quadrupedal and walk on all fours. But if a chimpanzee ambled up, it would stand underneath, reach up, and hang straight down from one of the horizontal bars. The upper body of apes, and of humans for that matter, is designed for hanging underneath branches. This mode of locomotion is called brachiation. For the ape to do this, the arms must be able to rotate 360 degrees in the shoulder socket, up to the ceiling and then down to the floor. With the shoulder blade on the back of the ribs rather than on the side, the ape arm is free to move directly overhead. Reach up, touch the ceiling, and feel the ability to brachiate. Apes can do this too, but monkeys can't. Some apes, including gibbons and siamangs, are singular brachiators; their major mode of transportation is swinging from tree to tree from underneath. Other apes, including chimpanzees and gorillas, are primarily knuckle-walkers. On the ground, this type of ape leans its upper body over its arms and rests its weight on the curl of the hands. In the trees, these large apes can brachiate, but sometimes the trees are too small to hold them, and ground travel is more efficient. Orangutans have the best of all possible worlds. They use grasping hands and feet to pass slowly through the trees and then knuckle-walk on the ground.

All apes are forest animals, but they differ radically in social structure and habit. Gibbons and siamangs live in the tropical jungles of southeast Asia. They're the penultimate arboreal primates. Black-bodied siamangs and buff or brown gibbons fly from tree to tree with the grace of airborne ballet dancers. They live in small family groups and mate in monogamous pairs. They maintain their way of life by loud calls that resonate though the jungle in the wee hours of dawn. Ohwaacka-ohwaacka-ohwaacka. Females call to females and males call to males. They map out family homes by mutual agreement (Mitani 1985a).

Orangutans live only in Sumatra and Borneo. These enormous red apes lead a solitary existence based on the position of fig trees in the forest. They move slowly, reproduce slowly, and rarely expend precious energy on social interactions (Rodman and Mitani 1987). The African apes—gorillas and chimpanzees—are more closely related to each other than to orangutans, and they're much more social. Gorillas, those in the lowlands and those in the mountains, stay in harem groups of one male with several females (Fossey 1983, Harcourt 1979, Stewart and Harcourt 1987). These gentle quiet creatures spend their lives digesting bamboo and avoiding poachers. Except for human interference, they have relatively peaceful lives punctuated by life-cycle events such as mating and births.

And then we come to the closest relative of all, the chimpanzee. There are two kinds of chimpanzees, the common chimpanzee and the less well known bonobo, sometimes called the pygmy chimpanzee; these parted ways from a common ancestor 1.5 million years ago. Some might say that today we really have three variants of that common ancestor; humans, chimpanzees, and bonobos. Most people know something about chimpanzees. For thirty years Jane Goodall has popularized chimpanzees in books, lectures, and TV shows (Goodall 1971, 1986). Her work, and that of her many colleagues, has provided reams of information on chimpanzee lives. Another site 150 kilometers south of Goodall's Gombe Stream, called Mahale Mountains, has also produce long-term data on chimpanzees (Nishida 1992). Chimpanzee populations range over west, central, and east Africa. Recent research on several populations shows that chimpanzees are actually very diverse in their behavior (Gibbons 1992). Although they all show a similar social system, and bonobos share that system, culturally they are quite different. In hunting practices, tool use, and some one-on-one behaviors, they differ according to their population (Small 1993).

Bonobos are less well known than chimpanzees for the simple fact that they are difficult to find. In the 1920s, bonobos were first identified as distinct from chimpanzees from skeletal material, but behavioral research has been conducted only sporadically since the 1970s. Bonobos now live on 13,500 square miles of forest in Zaire, a small patch of land in Central Africa looped by the Zaire River (Kano 1992, Susman 1984). Although Bonobos are often called pygmy chimpanzees, they are not pygmies in any sense. The average body size of the largest male bonobos easily overlap with the smallest chimpanzees, and females of the two species are the same size. Bonobos are, however, more delicate in build, and their arms and legs are long and slender—the welterweight of chimpanzees. On the ground, moving from fruit tree to fruit tree, bonobos often stand and walk on two legs—a behavior that makes them seem even more human than ape. But it's their faces that give them away as human cousins. Big brown eyes stare out of perfectly round, completely black faces. And many seem to have gotten up early and grabbed a comb just to make a perfect center part through the long back hair that flops right and left on their rather small heads, covering up their human-sized ears.

Chimpanzees and bonobos dwell in a social whirl. Groups split apart some days and come together others. Primatologists call this arrangement a fission-fusion society. There are also important differences between the two species. Female chimpanzees are rather solitary and tend to go off on their own, while males hang together like the boys in the back room. Chimpanzee males, who are often close relatives, probably brothers, defend the community's territory from other males who might want access to fertile females. Male chimpanzees are also extremely dominant over all females (Smuts and Smuts in press) and are surprisingly familiar with violence. For example, the chimpanzees at Jane Goodall's site in Tanzania waged war on a neighboring group of males, murdered them, and took over the new territory. When we use chimpanzees as models for our own behavior, we have to take the bad with the good and accept the dark side of our primate nature. The silver lining is bonobos. Bonobo males and females live much more egalitarian lives; males and females hang around together, and females are rather empowered in their comparative status with males. Female bonobos get priority at feeding sites, and there's no violence between the sexes. Bonobos spend much of their day in sexual interaction with one another. Although we can't claim either species as a direct ancestor, chimpanzees and bonobos teach us that complex social lives include anger, revenge, and murder as well as affection, attachment, sexuality, and affiliation.

The Common Ground

The trail from less familiar strepsirhine primates to better-known apes is littered with primates species with stories of their own. All animals are guided by what they eat, how they avoid predators, and the form and intensity of their social interactions. But they also have features in common, primateness that forms the core of their position in the animal kingdom. First and foremost, almost all primates, including humans, are social animals. Most primates live in groups, and by definition, group-living creatures have to get along. But primates do more than aggregate together to avoid predators. They interact on a minute-by-minute, day-to-day basis. Primates go to a lot of trouble to be more than social. Consider a group of wildebeests on the African savannah. They stand about, chewing grass, moving slowly across the plain. Sitting right next to the wildebeest group is a troop of baboons. The same few minutes of observation, concentrating on the baboons, would reveal a different social situation. The male baboons threaten one another with flashing white eyelids and open their mouths to display huge canines. Several female baboons groom one another, ever watchful of the dangerous males, lest they get caught in the squabble. Juveniles run about the troop, slapping one another and squealing. If one baboon reacts to the possibility of a predator, the group scrambles for cover. Females pick up babies, males grab juveniles, everybody screams and runs for the nearest tree. The quality and quantity of social interaction of the wildebeests and the baboons are strikingly different. One is an almost passive aggregation while the other is a seething pool of interpersonal interaction. This is a major primate feature, one we share with almost every species from lemurs to gorillas. (See Figure 1.)

The baboon example also suggests another major feature of primate lives—behavioral plasticity. Think of the family cat. Every time you come home, the cat greets you at the door and cries for food. You both trot into the kitchen and you lay out the kitty chow. But the cat isn't hungry: it sniffs the food and moves on to the next item in its ritualized agenda, maybe to go outside. It stands next to the door, you let it out, then kitty turns around and cries to come back inside. Day after day, you both repeat the same ritual. Sometimes the cat eats and sometimes it doesn't, but you're both hooked into this ritual of asking and feeding, regardless of what the cat's stomach tells it or what your brain tells you. Now imagine that the family pet is a monkey, let's say a capuchin. The monkey, too, would greet you at the door. But it wouldn't ask for food. Earlier in the day the monkey was hungry, so it ripped open all the kitchen cabinets, broke open a cereal box, and chowed down. The next day at monkey-designated mealtime it opened the fridge and stole an orange. When it wants to go out, the monkey finds out which window will budge and escapes all by itself. This comparison is not meant to belittle the intelligence of the domestic cat, only to demonstrate the difference between a small-brained animal and a large-brained one. The point is, the cat acts only in certain highly fixed ways; the cat's behavior is predictable. It can't reason beyond the "food ritual" and the "door outside ritual," and it rarely adapts its behavior to changing circumstances. The monkey acts and reacts. It changes its behavior according to the situation. Faced with the desire for food or fresh air, the monkey will solve the problem with what's at hand.

Primates also use their intellect to solve personal problems. Remember, most primates live in social groups and thus most of their difficulties are social dilemmas. Life in a complex group dictates keeping track of the social machinations all around you. Every move a social animal makes affects everyone else. Patterns of social behavior also weave a web of social interconnections among members of the group. Primates use a large part of their brain power to track social relationships. They watch one another and catalog what goes on. They also form alliances, coalitions, friendships, and use reciprocity to gain back favors. They even reconcile personal conflicts (de Waal 1990). A large brain serves as a computer disc for social interactions; it stores what others look like, what they do, and the play-by-play behaviors of everyone around. Some anthropologists have suggested that the superior social ability of primates is the driving force behind the evolution of the primate brain (Cheney, Seyfarth, and Smuts 1986, Small 1990a). This hypothesis is supported by the complex nature of primate social interactions and how important social networks are to this order. We might have big brains simply to keep track of everyone else's business.

Much of nonhuman primate behavior mirrors our own behavior, even to the untrained eye. For example, any human would see a connection between ourselves and other primates when an infant monkey joins a group. From the moment of birth, the mother is attentive. Troop members are fascinated, their natal attraction often a hindrance to adequate mothering. Yes, sometimes fathers care for infants, but this behavior is rare. In most cases, primate females are the important "other" in an infant's life. Chimpanzee mothers rear their infants for years, and even adult sons return for a maternal recharge. Macaque mothers reproduce almost every year and spend most of their adulthood nursing and guarding a series of infants. The intense mother-infant bond is clearly a major feature of most primates including humans. Or take for example the monkey attention to kin. Macaque relatives hang around together, eat together, groom one another more than nonrelatives do, and aid their kin in fights (Kaplan 1977, Kurland 1977). These broad patterns of behavior make evolutionary sense for nonhuman primates and they mirror our own behavior. We share these patterns of behavior because we are all primates that have evolved under similar evolutionary pressures and constraints.

What about mating? This book focuses on mating because it's the foundation of individual reproductive success. If patterns of mating have a biological basis, we would expect to see strong connections between the mating behavior of humans and other primates.

Primate females invest deeply in their infants and produce few offspring over a lifetime. This is the quality over quantity strategy—put all your eggs in one basket and watch that basket. In comparison, a species that favors quantity, such as the field mouse or the pig, produces many litters over a few years and spends minimal time mothering before beginning a new brood. Each strategy works, depending on the physiology of the female and the ecological conditions she must endure. Once the specific process has evolved, females must deal with the constraints of producing quality or quantity. For most primates, who don't produce litters (twins at the most), the process becomes a game of hard-ball—one egg, long gestation, months of lactation, and more months of care.

Theoretically, she shouldn't miss a step. She should be careful with those few eggs and mate with appropriate males. We might expect human females and their primate sisters to be selective, careful sexual partners. Like all primates, human females are interested in producing healthy offspring. A chimpanzee female is no more or less concerned about her infant than is a human mother. Are they both also concerned about the partners they choose to make those babies with? Humans and other female primates have so much basic biology in common, overall mating strategies of female primates should also stretch across the minor evolutionary gap between the species.

The system never works out perfectly and theory doesn't always match life. As the next few chapters will explain, our primate sisters

are much less careful when choosing mates than we might expect—
but then sometimes so are human females. If females' behavior differs
from what evolutionary theory predicts, it's only because the theory
hasn't yet taken into account some complexity of life, or perhaps
because the female sees a possible alternative. To understand the
various routes a female might choose during mating, we have to place
mating within the context of her life.

Clementine - 4 months

Baby's first leaf

Meeting Aunt Beckie

My 1st pregnancy

Perkins '92

CHAPTER 3

Enjoy Being a Primate Girl:
The Life History of a Female Primate

SHE WAS BORN May 22, 1987, and her name is Clementine. She arrived about seven o'clock in the morning after a very brief labor. Her mother caught Clem with her right hand, sliding the squiggling baby over her right leg and up onto her belly. Mother hugged the new infant as tightly as possible. Clementine blinked a few times and opened her dark brown eyes and stared into her mother's face. At the same time, baby Clem flailed both skinny arms toward a nipple that seemed oh-so-far away. Her mother was so exhausted by the time the placenta plopped out and hit the ground that she could barely move down the branches and out of the tree.

Clementine was the twenty-third infant born that spring into a troop of Barbary macaques, medium-sized monkeys native to Morocco. Her particular troop now lives in captivity, ranging over twenty acres of fenced oak forest in the southwest of France. The group of 150 animals has been under study by behavioral scientists, including myself, for several years. For the next few months, and periodically over many years, behaviorists will follow Clementine and her mother, write down everything they do, and compare that information with the behavior of other group members. These bits of numbers collected during hours and hours of observation paint a portrait of Clementine's life, what she does, and how that compares to the actions of other monkey females.

The object of this tedious collecting of data is to understand, by charting patterns of her behavior, what makes a primate female act the way she does. Given that Clementine, like all her primate sisters, is constrained by her biology, how will she run her life, and will she be a successful monkey in evolutionary terms? Will she choose the correct strategies that help her make healthy babies, bring them up well, and watch her children have children? Will her choices be the right choices in terms of reproductive success? A day in the life of Clementine reflects, in a broad sense, a day in the life of any female primate. I use her particular example to describe the life-history pattern of female primates. As a macaque, Clementine has species-specific options and constraints, but still she's a reasonably common monkey to take us through the primate female's life course.

Primates Studying Primates

When I took my first course in primatology in the late 1970s, I learned that female nonhuman primates grow up, have babies, and care for those babies. There was no mention of female competition, the social strength of female groups, or any kind of social behavior that wasn't somehow connected to motherhood. I heard what other budding primatologists, male and female, had heard for years before me. And before that, there was no such subject as primatology or animal behavior. As a science, primatology is young, and the abbreviated story of females only as mothers is a product more of missing data than of anything else.

Before the 1970s, primatologists didn't know very much about female primate behavior. The early work, mostly descriptions of one or another species, typically focused on the group as a whole or on males. Males are usually larger and more flashy, and they easily caught the attention of observers. Some suggest that there was also an unconscious antifemale bias during the early years of primatology—most of the observers were men and so they unconsciously focused on their own kind (Blaffer Hrdy and Williams 1983, Small 1984b, Wasser and Waterhouse 1983). Probably the bias was caused by a combination of factors: men's interest in male behavior, the Western perception that females were, and "should be," only mothers, and a lack of long-term information about what female nonhuman primates really did.

But there was a revolution afoot and it was heralded, strangely enough, by studies of females as mothers. Jane Goodall had, by this time, spent a decade watching chimpanzees at Gombe Stream in

Tanzania, East Africa (1971, 1986). Her main focus was mother and infant behavior, but along with her fine-detailed observations on mothers, she also gained a picture of female (and male) chimpanzee behavior in general. Even in her popular book *In the Shadow of Man*, Goodall describes female chimps as fully functional members of a highly complex social system which participate in activities other than mothering. At the same time, Goodall's description of female chimps turned out to be an unusual portrait of typical primate female lives because she was watching one of the few species in which females leave their natal group and end up rather solitary in unfamiliar territories. There's no real sisterhood among female chimps as there is in many other species. If Goodall has spent the same time watching baboons, for example, an entirely different picture of strong female bonds would have appeared. But Goodall wasn't privy to the elaborate sororal interactions of most social primates, simply because her species didn't behave that way.

Other women, less well known than Goodall but just as hard working, were also discovering the female primate within, and they happened to choose species in which females turned out to be the stable core of groups. The females of their chosen species don't leave at puberty, and for these primates, life revolves around female-female attachment. These women primatologists, including Sarah Blaffer Hrdy, Phyllis Dolhinow, Alison Jolly, Jane Lancaster, Suzanne Ripley, Thelma Rowell, and others discovered an array of roles for female primates (Blaffer Hrdy 1977, Jay 1965, Jolly 1966, Lancaster 1972, Ripley 1980, Rowell 1966, 1967). They revealed that female nonhuman primates were doing all sorts of things that have nothing to do with mothering. Females participate in intertroop encounters, they actively establish and maintain a hierarchy, and they often decide where and how fast a troop is moving.

Collecting behavioral work was a rather slap-dash affair during those early years. A person found a group of primates, tried to get them accustomed to human observers, and wrote down everything they did. There weren't any university courses in primatology, as there are today, nor were there established methods for collecting and analyzing behavioral data. These early studies were like the work done by cultural anthropologists in the 1930s and 1940s whose mission was to canvass the globe in search of aboriginal peoples and record their cultures before they disappeared. Early primate researchers believed then, and believe even more so today, that the forests were rapidly shrinking and that it was imperative to gain a general picture of each species in its habitat before it was wiped out. But by the mid–1970s general descriptions of primate groups were not enough, because a revolution had occurred in the study of animal behavior.

The change in direction was prompted first and foremost by a move to use an evolutionary framework to explain why animals behaved in certain ways. In their search for understanding how Nature operates, scientists had begun to realize the importance of behavior to reproductive success. Evolution could explain why animals look the way they do, and it could probably explain why they act the way they do. Evolutionary biologists interested in behavior began with the simple premise that each individual is born to create other individuals. In other words, individuals "should" act selfishly when their behavior influences survival and reproduction. Those who act in appropriately selfish ways will pass on more genes. The others will be reproductive losers. Individuals could also behave "altruistically" toward kin and thus pass on some of their shared genes if the costs were not too great (Hamilton 1964). Through kin selection, the favoring of kin over nonkin in some social interactions, their genes live on, even if the individual dies. Animal behaviorists gained an evolutionary framework, a simple one at that, on which to hang an animal's behavior. The framework became official when the discipline gained a name in 1975. In his comprehensive volume *Sociobiology*, Edward O. Wilson, a Harvard entomologist, laid out the blueprints for understanding social behavior, both for human and nonhuman animals (1975). The ground rules for any species are evolutionarily logical: behave in tactical ways that help you survive, mate, and produce viable offspring and you win the game of reproductive success relative to others. Now that the framing was in place, it was up to behaviorists to test what animals were doing. Thus was born the science of behavioral biology. Today we know how female primates run their lives and why some female tactics evolved. The information has come from many species and from long-term field studies, from shorter concentrated research, and from laboratory experiments that test certain tightly constructed hypotheses. Primatologists have looked at female mating behavior, aggression, competition, friendship, and old age.

The rapid increase in studies of female primates in particular was also aided by an important sociological phenomena. As in most sciences, what people choose to study is highly influenced by their milieu. A person's gender, upbringing, and outlook on life have a major impact on the choice of scholarly pursuits. The feminist revolution in the United States in the 1970s, and less so worldwide, echoed what was happening in primatology. More women entered graduate school, more women went into the field to study animals, and more women turned to primates to figure out womankind's place in Nature. Yes, many men, including James Loy (1971), Jeff Kurland (1977), and others, were instrumental in increasing our knowledge of female primates, but women had an inordinate influence on how female nonhuman primates were studied (Blaffer Hrdy and Williams 1983, Small 1984b).

This rich heritage of research allows me to describe with confidence Clementine's early life as a macaque, predict how her life will proceed, and provide explanations for why her life might take specific directions.

Clementine's First Day

The first day of Clementine's life was a microcosm of her life to come; no fortuneteller could have shown her a better picture. From the moment of Clem's first peep, which was a kind of self-generated birth announcement, her whole troop was alerted to her presence— no quiet birthing room here, no notification of relatives first. Her high-pitched squeak coupled with the flamingo pink skin of a newborn monkey served like a magnet for the other group members. They all turned their heads to stare at her mother. A new member had joined the troop and every monkey seemed compelled to find out just exactly what was wobbling about there. Some ventured close and tried to touch her. One adult male even reached out and grabbed her arm, but Mom turned her back and walked away carrying Clem. She wasn't quite ready for all the attention this infant would receive; just a few days to be alone with her baby, please. Compelled by some primal urge to bond with Clementine, her mother moved a few feet away from prying eyes and tended to her infant.

This story of Clem's mother and her need for a strong mother-infant bond is familiar to us humans. The close relationship between an infant and its primary caretaker is, in fact, one trait that distinguishes primates from other mammals. If you count the number of mammals and subtract all those producing fast-matruing infants who quickly fend for themselves, only primates, with their never-ending need for parental care, will be left. Yes, most babies need some care, and yes, all mammal infants are nursed, but what distinguishes primates is the intense quality and long duration of the mother-infant relationship. It's always more than a few days, most often many months, and in the case of humans and chimpanzees, will last for a lifetime. This bond is fostered both by the physical needs of the infant—it must be nursed and carried or it will die—and also by a social and emotional bonding that turns out to be just as important for primates.

In the 1960s, Harry Harlow, a psychologist, conducted a series of experiments on captive rhesus monkeys which would revolutionize how we think about human behavior, especially maternal behavior. Harlow was interested in the social interaction between mother and infant and how that bond developed (Harlow and Harlow 1965). In his most influential experiments, Harlow separated newborn female monkeys from their mothers and placed each of them in a cage with two fake, human-made, mothers. The baby monkey had a choice. It could spend time on a wire-mesh model of a mother with a plastic moon face and warm milk flowing from its rubber breasts. Or the infant monkey could hang onto the other substitute, a similarly shaped wire mother wrapped in terry-cloth toweling. All the rhesus infants tested did the same thing—they went to the nursing wire mother many times a day to suck milk, but they never spent any time with her. Instead, all the infants clung tightly to the textured warmth of the terry-cloth mother. Harlow interpreted their behavior as proof that primate infants need maternal connections even more than they need food; touching, feeling, and holding are important parts of the bonding process for us primates. But his work told us more about why we primates need one another. Many years later, when these motherless monkeys grew up, they were inadequate mothers themselves. They were unable to interact with other members of their species, even to mate. When they did conceive and bear infants, they had absolutely no knowledge about caring for them. They were socially dysfunctional. In evolutionary terms, the motherless mothers would be failures because they couldn't help their own offspring reach maturity. These experiments suggest that attachment to a mother or to some primary, living, breathing other is vital to normal social development for primates. Without this primary attachment, we primates are truly social zeros.

This news is no surprise to most primate mothers. Clementine's mother, with her swift move away from an annoying male, was only obeying her instinct to protect and nurture her new addition. A voice inside her little monkey brain said, "Keep this thing close, guard it from harm, let it eat, it's yours." This is the genetic program that triggers the initiation of the mother-infant bond at the moment of birth. If initiated and reinforced, the bond will work correctly and Clementine will follow a path trod by millions of other female monkeys before her. She'll learn the appropriate macaque social rules as a juvenile, make mate choices once she reaches adulthood, treat her infants well, and grow into a matriarch. If the bond is shattered by her mother's death, or if, for some reason, she skips some of the lessons about being a monkey, she'll fail in one of her many life steps. In evolutionary terms, her life will be a waste. The successful female monkey balances her constraints with her opportunities and moves into the next life stage. And how she does it isn't exactly left up to chance.

The first few weeks of any primate's life are a social whirl. Consider the birth of a human baby. First the family is notified, then the larger circle of friends. Day after day, the infant is introduced to those who will be part of its life. And more remarkable, every person even remotely involved with the parents is curious about the new member of their species. Bring a brand new baby into an office building and watch all work come to a standstill. Try to pass a carriage without looking inside. What sex is it? What's its name? Right off the bat, other humans want to classify this fellow member of the species and catalog it into some mental social register. Other primates do the same. They sniff, touch, lick, and stare. They slide up close and make faces at the newborn. They're just plain curious. For the infant, this is a time of curiosity too. Primate infants, be they lemurs or gorillas, spend most of their day staring at others. Because they ride on mothers (or sometimes fathers), infants have a particularly good seat when it comes to "people watching." Infants spend most of the day integrated directly into the network of mother's friends and relatives. The first circle of social connection is dictated by mother's social register and, if it's a matrilineal social structure, blood relatives. The social system therefore perpetuates itself.

At this life stage, there isn't much difference between a male infant and a female infant in their daily habits: both genders suckle, sleep, squeak at others, and ride around that at similar rates. Some researchers have found mothers treat their infants differently according to the infant's sex, rejecting males and encouraging independence sooner in males than in females (Berman 1984). But most often, mothers treat their infants the same regardless of gender (Nicholson 1987). Do monkey mothers sometimes "know" the sex of their infants? Presumably they do. It isn't much to expect a monkey, or an ape for that matter, to catalog its troop mates into two distinct categories, male and female. It's only reacting to the distinct patterns of behavior exhibited by each gender, whether or not it has names for the two sexes. Monkeys certainly know which group they each belong to when they exhibit typical "male" or "female" behaviors.

Baby Needs a New Pair of Shoes

At some point, baby primates initiate independence from their mothers. The tiny infant is still wobbly, unsure of its own locomotor pattern and more sure of its mother's steady body. Independence is characterized by time spent off its mother, feeding alone on foods other than mother's milk, social independence, and, most important, play (Nicholson 1987).

Play is critical because it guides social learning. For Clementine, the transition to independence, begun at day one when others tried to grab her from her mother and completed some seven months later when she spent most of her daylight hours off mother, was facilitated by her same-age social group. Because Barbary macaques breed seasonally in the fall and infants are born in the spring, Clementine had a cohort of ready-made pals her same age. Especially important were those infants born to her mother's sisters and friends. Other primate young have a more difficult time with this life stage. Take, for example, orangutans. Their mothers live very solitary lives. Home ranges of females barely overlap, and social interaction occurs almost by accident. When a baby orang wants to play, it has only its mother and perhaps an older sibling as a partner. Play, a universal feature of all young animals, is a rather restrained affair among the solitary orangs. Clementine, however, had not only her same-age peers but lots of one- and two-year-olds to harass. The difference between Clem, the playful macaque, and the more restrained infant orang also reflects the role of social interaction in their respective adult lives. Through play, each young primate learns how to interact with others, and in the orang case, social interaction is not as important as it is for the macaque.

One of the joking rules of primatology is, never take data on juveniles. They look too much alike, move too quickly, and spend all their time playing. An observer needs a slow-motion video to figure out exactly what's going on from minute to minute. Within the context of play, each primate gains, even at this young age, an appreciation of relatedness among group members, its place in the hierarchy and the role of status, and an appreciation of its sex role. Females actually play differently than males do (Walters 1987). In species that live in highly complex social groups, males play more frequently and engage in rough-and-tumble play more often than females do (Caine and Mitchell 1979). This sexual difference in play appears in all the studies that have been conducted on juveniles, including macaques, baboons, and langurs (Walters 1987). The difference is apparently hormonally based. When female rhesus monkey fetuses given male androgen hormones became juveniles they played more like males than females juveniles (Goy and Resko 1972). But the difference also makes social sense because males and females will follow separate life history patterns into adulthood. As adults, the males will need to fight with other males to gain access into a new troop when they emigrate and try to obtain mates. Female juveniles, in contrast, tend to stay close to home, interacting most often with the daughters of relatives. At this young age, females are forming bonds with other females, making friendships, and learning their place in the female hierarchy.

Juvenile females also differ from males in social behaviors that echo their future role in the group. Most people who watch monkeys in a zoo cleaning bugs from their cage mates think that grooming is an idle exercise. But grooming is anything but a casual form of hygiene. It does remove unwanted parasites and flaky bits of skin, but long years of watching primates have demonstrated that grooming, the passage of hand and fingers though another's fur, is a mark of attachment. Clementine and her mother groomed each other more than they groomed others. As a youngster, Clem also groomed high-ranking females more than low-ranking females. A low-ranking female, Becky, tried to worm her way into Clem's heart by grooming her for long hours. In every case, the physical touching, the number of times grooming was performed, and the length of each grooming bout were indications of an established relationship, or one in the making. Just as we touch and sit close to those we care for, monkeys and apes groom their special "others." It follows that grooming will be performed most by individuals who want to foster relationships, and those individuals are most often females. Studies of macaques (Goy and Resko 1972, Kurland 1977, Sade 1965, Silk, Samuels, and Rodman 1981, Yamada 1963), of vervets (Fairbanks and McGuire 1985), and of baboons (Cheney 1978, Walters 1987)—all group-living species wherein females stay home and live by interpersonal interactions—show that female juveniles spend more time grooming than do males. They also initiate friendships with their willingness to groom at the drop of a hat. But there's method to a female juvenile's subservient ever-ready grooming behavior. Primatologists have noted that juvenile females most often attempt to groom those higher in rank when there's no blood relationship between them, a strategy they'll follow into adulthood (Walters 1987). In other words, schmoosing begins early with a flick of the wrist. And for females, a grooming policy later in life may allow them to get close to preferred males, calm high-strung relatives during a fight, reconcile after a nasty conflict with matriarchal relatives, or keep infants by their sides. Grooming is a female's primary social tool, and she learns to wield it early.

The Babysitters' Club

The young female also uses the juvenile period to practice her future role as mother. When a new infant is born, often the first curious face an infant sees is its older sister's. In Clementine's case, this was also true. Her mother had given birth to another female two years previously. This little female wasn't yet sexually mature and she still played with her youthful play group. But she also spent much of her time with her mother, always sleeping with Mom at night. The new baby was an intrusion into older sister's time with her mother, but the presence of a new sibling also afforded the opportunity to interact with an infant. Propelled by some primal urge to hug an infant, the sister was the first monkey besides Clem's mother to smell, touch, and hold the new infant. This pattern of juvenile interest in infants is a widespread but not a universal feature among primates (Caine and Mitchell 1980, McKenna 1979, Quiatt 1979). Jeff Walters calculated that juvenile female baboons interact with each infant in the group at an average of once every hour (1987), and Nancy Caine discovered that rhesus juvenile females spend almost four minutes every hour with infants (Caine and Mitchell 1980). Studies of other monkeys show that these little females don't much care which infant they get, as long as they have an opportunity to touch and grab something that squiggles. Mothers, however, aren't quite sure that they should let juveniles drag their infants around. In fact, low-ranking infants are often harmed by intense juvenile (or adult for that matter) attention (Silk 1980). Thus, most often, anxious little females are allowed only limited time with a sibling, and if this brother or sister screams, the mother quickly grabs it away. Primatologists such as Jane Lancaster (1972), who first wrote about the juvenile female love affair with babies, have suggested a potentially simple explanation for this phenomenon. Young females have been preprogrammed to be attracted to infants because this attachment triggers a mothering response. Without this initial push and subsequent practice, they suggest, female primates would have no clue about parenting. This hypothesis has some problems, however. Many young females ignore babies or have no opportunities for practice because so few infants are born, and yet they grow up to be fine mothers. Thus the direct connection between early babysitting and later maternal expertise is less clear than might be expected.

First Love

The juvenile stage comes to an end as the young female primate reaches sexual maturity. In human females, menarchy, or the onset of menstrual cycles, signals the beginning of the ability to conceive, and this is usually a girl's first step to womanhood. A young woman experiences changes in her body shape with the addition of breasts, pubic hair, and curving hips. She'll also see changes in her social status as she falls in and out of social cliques, always yearning for high status and "fitting in."

For other primates, the gateway to adulthood is very similar. In the spring of Clementine's third year she played less than during her second year. She also gained some weight—an adolescent growth spurt—which prepared her body for the demands of reproduction. She spent her social time both with her mother and with friends she had established during her juvenile months.

When breeding season began in the fall, three-and-a-half-year-old Clementine experienced her first cycle, and this was the clearest signal of her transition into adulthood. Some nonhuman primates breed only during particular sequential months called a breeding or mating season. When seasonal breeding is characteristic of a species, it seems to follow seasonal fluctuations in resources important for the primates involved (Lindburg 1987). For example, the mating season often occurs during the rainy season so that infants will be born when leaves are budding and the demand of lactation can be met. Female cycles, whether they occur synchronously during a defined season or year round according to an individual female's life course, are called estrous cycles rather than menstrual cycles. This is the period of "heat." Only during these cycles are females interested in sexual behavior, and it's only during this time that males are interested in them. We know conclusively that in twenty species of primates out of two hundred, females signal their sexual readiness with temporary changes in their bodies such as large swellings on their rear ends, olfactory cues, or behavioral gyrations (Blaffer Hrdy and Whitten 1987).

Think of the chimpanzee female you've seen at the zoo. Her backside doesn't look right—in fact it looks as though she has some sort of large pink tumor growing on her butt. The area around her anus and her vaginal opening is swollen beyond recognition—at least beyond human recognition. For a male chimpanzee, this huge pink swelling is a beacon in the storm. It means the female is reproductively active, ready to mate. We also know that it means she's close to ovulation. The hormonal changes in the female chimp's body have changed both how she looks and how she acts. Most of the time she will run from males, even if they are interested. But once the hormones of reproduction kick in, she willingly moves toward a persuasive male (Goodall 1986, Tutin 1979). She backs up to the male, crouches down, and facilitates the insertion of his penis; she wouldn't be caught dead doing this any other time of her life. For most—but not all—female primates, sex has a time and a place.

Clementine's behavior as a sex partner was complicated by her age. Because this was her first heat, she cycled later in the season than did older females. By the time she experienced her first cycle in December, most of the other females had cycled for three months and were already pregnant. In addition, Clem's first cycle was probably infertile. Adolescent females always have a period of sterility during which they show all the outward signs of cycling, but no healthy egg is produced. Linda Scott, a primatologist who studied young female baboons for two years in Africa, found that males treat young females and fully adult females differently (1984). She was observing a group of olive baboons, a species that lives in large social groups of over a hundred animals. Females stay in their natal groups as they grow up, and like Clem, they find a place in the female hierarchy. Males, on the other hand, drift away from the group at sexual maturity, and transfer to new groups, often switching groups several times before they settle down. There's no breeding season among olive baboons; females are on individual schedules of cycling, pregnancy, and lactation. At any given time, one or a few females will be in estrus, sporting large swellings and causing competition among males for her favors. Scott discovered that males virtually ignored young females for several cycles, not wasting their energy or sperm on females who might not conceive. Little females, however, were persistent, constantly soliciting males; they just weren't successful in gaining males for several cycles. Scott believes that the males were responding to a lack of some sort of cue that females in adolescent sterility don't have. Males aren't interested because the young females aren't really fertile yet.

Travelin' Gals

This portrait of growing up as a monkey is actually rather one-sided, focusing as it does on macaques and baboons. Seen though their glasses, the life of a maturing female appears rather serene. Stay with your mother and sister, reach puberty, mate, and get on with life. Richard Wrangham has called this system "female-bonded" because females remain together and their most important relationships are with female relatives (1980). But several primate species don't conform to a female-bonded description, and for those females, life is quite different. Unfortunately, the overwhelming majority of field studies have been done on semiterrestrial species that are female bonded (Southwick and Smith 1986, Strier 1990), and thus we tend to think of all female primates as bonded, but this just isn't so. Gorillas, for example, the largest primate and one of the four apes, operate under a system markedly different from that of macaques and baboons—different in that both females and males leave the group at sexual maturity. Other species of primates in which females emigrate include chimpanzees, South American howler monkeys, and woolly spider monkeys. Given their smaller size, why would females leave the safe haven of a natal troop and venture into the unknown?

Karen Strier, a primatologist, who has watched female woolly spider monkeys leave their natal troop repeatedly over the past fifteen years of her field work, suggests that we're asking the wrong question. Because female emigration is found across so many genera, she points out, it may be a primitive trait. From this perspective, females staying with their natal group would then be the later-derived pattern that evolved under some sort of special selection pressure (1990). In that case, female bonding, not female emigration, requires a novel explanantion. But others insist that female emigration, just because it's more unusual, should be explained. Various hypotheses have been offered to account for the roving urge in some female primates. The easiest is that someone has to leave because of inbreeding, and if males don't go, females have to (Moore and Ali 1984). This explanation may satisfy the case for chimpanzees, whose males quickly band together to defend their territory against intruders. The best bonds are made with kin, and these chimps are probably brothers. Females then are theoretically forced to leave the area to find unrelated males. Gorillas live in fairly small groups of five to fifteen animals, dominated by the silverback male. This male fathers all the infants, and when a female grows up, she must leave the group to find a mate other than her father (Stewart and Harcourt 1987, Watts 1991). Her large body size, still half the weight of males but larger than anything else in the forest, gives her safe passage to the next group. Interestingly enough, female gorillas transfer more than once. They move out of their natal group into another group, produce one infant in that group, and then move on to another. David Watts suggests that what motivates a female is the search for a male who will protect infants from infanticide. It seems that the male attack on infants is a major reason for infant mortaility among gorillas, and thus a female is always looking for that special male who, in protecting her from other males, will also protect her infant (1991).

The case of howler monkeys is more difficult to explain. For these loud group-living animals, there's no clear reason why females leave. Just like males in most multimale-multifemale groups, male howlers move on to new groups once they reach maturity. But oddly enough, females also sometimes feel compelled to move on. Carolyn Crockett studied female red howlers for years and noted that the identity of moving females had something in common—most of them were daughters of low-ranking mothers. Crockett suggests that the desertion by some females is a result of breeding competition. If these low-ranking females stayed in the group, they'd be harassed by higher-ranking females, never given access to males, and perhaps starve to death (1984). Females compete with one another for a certain number of breeding spaces, units defined by the habitat and the availability of food. Once the howler's world is full, someone must move on. Thus the strategy of these wayward females might prove to be the best one. If they are allowed into another group, they might have a chance to survive and reproduce.

In any case, the female's desire to move away from home makes these groups very different from those of macaques and baboons in another way. Once a female leaves, her bonds with family are broken. She can't rely on mother, sisters, or any members of a potential matriline. Her bonds, as they do for gorilla females, will center on males rather than females, and she must build a nonkin sisterhood all by herself.

Dominance

I remember sitting in my first primatology class, hearing a lecture on dominance hierarchies, and thinking, "Sure, the professor is exaggerating. Only humans are so interested in status." A year later I was watching a group of bonnet macaques as part of a study of female behavior, and after my first hour I realized that the professor was right. If you spend more than a few minutes with a group of macaques and focus on the patterns of interaction of a few individuals, a consistent pattern among them will eventually emerge. Monkey A does a behavior that's clearly dominant to monkey B, even to human eyes, and B will respond in a submissive way. If you watch long enough, and write down these behaviors, you will be able to line up their relationships like a stack of playing cards, each one more valuable than the one underneath it.

Much of the earliest work on female primate behavior which veered away from motherhood concentrated on female rank. Rank was important for the monkeys; thus it became important for those watching them. Primatologists learned quite early that females of high rank mature earlier, have more infants, and most often have infants that reach sexual maturity (Fedigan 1983). In other words, there's a positive correlation between female rank and reproductive success for those groups in which rank plays a clear role in intermonkey interactions. How exactly does this work? It turns out that just like males, females of high rank are privileged animals; they simply get more to eat and this provision eventually translates into healthy offspring. I discovered that even in captivity, where there's plenty of food and it's evenly distributed, high-ranking females are fatter than low-ranking females (Small 1981). Patricia Whitten looked at the effect of rank on the feeding habits of vervet monkeys in northern Kenya. She watched females as they ate, writing down the type of resource and noting fine details such as the ripeness of each item a monkey placed in its mouth. Vervet females mostly eat fruit, flowers, seeds, and thorns, and these resources differ in an important ecological element—distribution. When a food resource is clumped, such as the flowers of an acacia tree which vervets prefer, high-ranking female get more. Even more significant, the intake of flowers from another species of acacia is apparently associated with the early onset of breeding season; in this case, high-ranking females who have priority of access to these flowers will breed sooner. High rank and the privilege of eating the special flowers might then have an overall effect on the reproductive success of the high-ranking females (Whitten 1983).

Dominance hierarchies function in monkey society just as they do in human society; those higher up get more. High-ranking primates gain more food, or most of the preferred food, and they are often the winners in the game of reproductive success (Fedigan

1983). Perhaps the most dramatic lesson in the importance of rank to female health and survival is provided by the toque macaque found on the island of Sri Lanka. Several groups of this small gray-brown monkey have been studied for over twenty years by Wolfgang Dittus, a primatologist with the Smithsonian Institution (1977, 1979). The toque monkey, closely related to its Indian cousin the bonnet macaque, is a comical figure distinguished by a spiral of hair sitting on its head like a halo gone flat. The monkeys live in large social groups, and typical of all macaques, males emigrate at adolescence. Females form large, closely linked matrilines. During Dittus's study, Sri Lanka experienced a major drought, which decreased the available vegetation for the monkeys. Dittus discovered that mortality was also high during this period, and juvenile females suffered the most. The group, in general, became hostile. They supplanted one another over food right down the line—all dominants over subordinates, males over females, females over juveniles, juveniles over infants. High-ranking animals were observed literally taking food out of the mouths and cheek pouches of lower-ranking animals. But juvenile females were exploited the most, and their high mortality can be explained, in part, by their failure to thrive in this oppressive situation (1977, 1979).

Dominance rank is an important feature of many but not all primate groups, and as juvenile monkeys learn about their society, they also learn their rank order according to their gender: little males relative to all males and little females relative to all females. Most cercopithecines—Old World monkeys including baboons, vervets, and macaques—run their lives through a linear matrix of dominant/subordinate relationships. Because Clem is a macaque, her life too will be organized around this rigid maze of interpersonal dyadic interactions. I know this because in over six hundred hours of watching her group, I saw 1,126 clear hierarchical interactions among the forty-two adult females—an average of about two interactions per hour—and these were only the ones I saw right in front of me. Thus the rate I saw gives only a conservative picture of the daily intrigue of this hierarchical society. The ranks of adult females, at least among macaques and baboons, are rather stable. There are changes as more females enter the hierarchy, and sometimes those at the bottom will rise (Small 1990e), but compared to the ranks of males, Clementine's position will remain relatively stable for most of her life.

I use the word "stable" with caution. It sounds as if each female monkey moves into a status parking space at adulthood and turns off her engine, whiling away the days until old age. But the hierarchy of female monkeys is kept stable only because it is constantly rein-forced. Clementine's place in the female hierarchy was still low because she had only just begun to challenge the older adult females who ranked below her mother. Over the course of the next two years, she would rise in rank to just below her mother. And in an odd twist of sisterhood, she would rank above her older sister. The human observers noted several subtle cues in her hierarchical climb. One morning she walked past an older higher-ranking female, who quickly moved out of Clem's way. Another afternoon she bared her teeth at an even higher-ranking female and the target of her disdain cowered. Added together, these one-on-one interactions painted a picture of Clem's social power in the group. She was often aided by her mother and her older sister, and if the fighting got too rough, her mother's friends. Every day, almost every hour, Clem and other young females tested the waters of social interaction to find their rightful places. And like the positions of humans who quickly gain or lose a fortune, her status can drastically change if her circum-stances change. If the group fissions, if the high-ranking matriline experiences a number of deaths, if an up-and-coming matriline bands together and launches a coup, or if any demographic change befalls a population, the ranking of this troop will be affected. Throughout her life, however, Clementine's status, and the rank of her whole matriline, will dictate most of her interactions with other females and her access to food. She will rely on her mothers, sisters, and cousins to back her up during a fight. And along with her matrilinial members, she'll be able to displace others from important food resources. This sisterhood is important for female primates because it helps them remain a strong cohesive social power, and lack of this power can have drastic consequences. Clementine began her ascent into the adult female hierarchy during her first breeding season. She'd learned a lot about social interactions during her juvenile years. She knew how to intimidate others with an open-mouth threat, canines exposed and a sharp hiss escaping from her throat. She could also easily walk up to some lower-ranking females, and her mere approach would cause them to move. But these dominant behaviors were also directed toward her by females higher ranking than her mother, stopping her climb to the top.

For macaque males, who come and go from the more stable core of females, the hierarchy tends to be less tight than the female hierarchy, but adult males in general rank over all females. Males also acquire their ranks differently. When a male macaque transfers into a new group, he begins at the bottom and must work his way up. Presumably size and aggressive skill inherited from his parents will come in handy. His social acumen should also help ingratiate the new male into the hearts of females (Berard et al. ms.). For females in matrilineal groups, however, rank is most often inherited in a direct manner. It's based either on age, as it is in Indian langur monkeys (Blaffer Hrdy and Hrdy 1976), or on the mother's status. In Clem's group, and in all groups of macaques and baboons, the latter kind of hierarchy was established among females eons ago and is passed down through generations. If the mother is high ranking, so will her daughters be. If low status is the mother's fate, her daughters have little chance of moving up the hierarchy. Juvenile daughters of low-ranking mothers move away from high rankers with the same jumpy fear as their mothers'. The system is thus perpetuated.

Sisterhood Is Powerful

The adulthood of most female primates passes though systematic stages of cycling, pregnancy, lactation, and child care. There isn't much down time for a female primate without birth control. But life isn't all mothering, and an attempt to shove primate females, be they apes or humans, into a mother-only role is incorrect. As Sarah Blaffer Hrdy has pointed out, "Every female is essentially a competitive, strategizing creature" (1981, p. 97). Just like males, females must stay alive, avoid predators, and get enough to eat. They also must find time to bring up babies.

For females of many species, including forest-dwelling monkeys, rank is of no particular importance. They form their female-female attachments through grooming, sitting together, and touching. Rank-oriented females also perform this grooming because relationships are not built or broken on intimidation alone. Clem will spend much of her adult time grooming her own offspring and her other kin.

There will be touching and sitting close, and all these behaviors seem positive. But keep in mind the evolutionary basis for forming relationships. Because these animals have been selected to live as social, group-living individuals, all interactions have been selected over time as useful in the long term. Evolution doesn't favor something because it's nice. Selection favors behaviors that help animals stay alive and reproduce. In other words, females are strategizing creatures, always on the lookout for ways to improve their lot. Females band together when grouping suits their own needs, and they favor their kin because kin can be trusted (see Figure 2). If a female is a member of an interactive social community, she will use her social skills to achieve rank, maintain rank, and use rank to gain special foods—maybe even mates.

Clementine will also form adult friendships based on a tit-for-tat basis. Following on the heels of understanding the importance of kin relationships among primates came the odd observation that these animals also form relationships that have nothing to do with kin. Robert Trivers, an evolutionary biologist, suggests these relationships are formed and maintained by reciprocity (1971). Imagine this scene— Clem is in the middle of a fight. She screeches to her sisters, who come running, but their help is not enough. She runs into the core of the group and bows low to one of the the highest-ranking females, Maude. Maude quickly responds by rushing Clem's opponent and the fight breaks up. Why would Maude bother to come to Clem's aid? If we were to trace the history of the relationship between these two females, we would see a long-standing "I'll scratch your back if you scratch mine" relationship. It turns out that, this high-ranking female is the same age as Clem; they grew up together. They had formed a friendship long ago and maintained it by mutual grooming and support. And if we spend enough hours watching either of these females after the current fight, we would certainly see a payback on Clem's part. It might come as a long bout of grooming or as support to Maude in a future fight. But it would come.

Consider the last gift you received. Didn't you immediately remember that you had previously given the giver a gift? And if you hadn't been generous in the past, you probably make a quick mental calculation determining how soon you're required to return the favor. Several times a day, if you are lucky enough to have many friends, little favors are given and received, and the best friends are those with whom we have a balanced reciprocal relationship. Some might be amazed that mere monkeys are keeping such close track of their friendships. Humans do this easily, but lower primates? Although nonhuman primates may not give and receive material goods, they are quite adept at tracking social relationships. In fact, the reason primates, including humans, have such big brains may be to accommodate all these interpersonal scores (Cheney, Seyfarth, and Smuts 1986, Small 1990a). We've yet to prove a direct relationship between social acumen or reciprocal friendships and greater reproductive success, but no primatologist who has spent more than a few minutes watching any prosimian, monkey, or ape group would deny the importance of nonkin-based relationships to these animals.

Working within the (Mating) System

Each species of primate has evolved a particular social system and a particular mating system. The appearance of any particular system is the result of the genetic history of a species and its ecological constraints. Resources are the key to the distribution of female primates relative to every other female. It might be to a female's advantage to live, feed, and reproduce alone, assuming she can occasionally meet a male. That way, she has no competitors. But she might also need other individuals to defend a territory or find food. In that case, her best bet would be to stay with female relatives. The male's problem is not really food, it's fertile females. Even if he gets enough to eat, he also has to find females before he has even a chance for reproductive success. Thus males will be compelled to try and sequester as many females as possible. Years ago, the primatologist Richard Wrangham suggested that as a general rule, a social and mating system evolves first because of the needs of females (1980). Female distribution is dependent on resources—they may be close together or spread out, in small groups or large groups. And then males are layered on top of the female distribution. The result might be a monogamous pair, a small harem, or a large group with many males and females.

Solitary animals have a social system that doesn't allow much sexual philandering. African galagos, Asian lorises, and the orangutans are basically solitary animals. But when mating time comes, they somehow find a mate. Galagos and orangutans have evolved a similar solution to their mating problem. Females occupy small territories, which they share with their young offspring. One resident male lives in a much larger territory that encompasses the spaces of several solitary females. The male monitors "his" females for signs of estrus, and he defends them against intruding males. In a sense, this is a wide-ranging harem in which the females rarely meet.

As I mentioned before, there are more monogamous species among primates than among any other mammalian order. Only birds have more monogamous groups. Monogamy is a relative term for primates, however, as I discuss later. Researchers have recently discovered that many primates living in pair-bonded social systems don't always remain faithful to their partners (Palombit 1992): both males and females often copulate with their neighbors. Still, in a monogamous system, finding mates is facilitated by the fact that a partner *is* potentially always available.

There's only one polyandrous primate, one species in which a female has more than one male all to herself. The tiny saddleback tamarin of South America was first thought to be monogamous, like all other tamarins and marmosets. Researchers working with saddlebacks in captivity always placed a male and a female together, and when they introduced another female—a test of polygyny—the first animals went crazy. But no one thought to introduce an additional male. In her field work in Peru on saddlebacks, however, Ann Wilson Goldizen noticed something odd (Terborgh and Wilson Goldizen 1985). In the majority of her supposedly monogamous "pairs," there was an extra male. This fully adult male shared care of the offspring, and it was impossible to determine if one male was more the mate than the other. Saddleback tamarin females share copulations and parental care with two males.

Some primates live in a harem social system. In this case, we assume that a male is able to keep other males away from a small group of females who are attached to a fairly small patch of territory. Harem males tend to be much larger than females because evolution has selected them to be large to fight off other males. Gorillas, patas monkeys, redtails, black and white colobus, and some langurs live in harems (Cords 1984). But females living in harem systems with one male may also have opportunities to mate outside the system—their mating system may be a little different from their social system. Gorilla females have it easiest. Many gorilla groups (about 40 percent) are not strictly harems. The six or seven females are indeed guided and dominated by a huge silverback, but another male often sits quietly among the group. He's called the blackback because his hair hasn't yet turned the paternal silver of older males. He's tolerated by the chief, and may even be a brother or son, and best of all from the young male's point of view, the blackback also gets to mate. What appears as a harem is really a two-male opportunity for females. In addition, gorilla females leave their groups at least once and move into new groups, attracted by breeding males (Stewart and Harcourt 1987).

The many male–many female system is the most common among primates. Males have access to any female in estrous, and they may have to compete among themselves for her favors. As a result, males in this type of system are usually larger than females, a response to male-male competition. At the same time, females also have access to many males—a system that primatologists awkwardly call po-lygyandry. Clementine's mating will occur within this multiple-mate system, just as do the matings of all macaques, baboons, vervet monkeys, and many others. Once she reaches reproductive age, any number of males will be interested in her when she comes into estrus. With her huge swelling, all males will be attracted, and no single male will be able to sequester her. The details of the multiple-mate system change according to variables such as seasonality or the visible signs of estrus. But the key here is that females have more than one male to choose from.

Mating, Gestating, and Lactating

Nowhere is strategizing more apparent than in the mating behavior of adult females. Covered in more detail in Chapter 6, the mating choices of females, and the options open to them, paint a clear picture of females guiding their own reproductive ship. They're neither coy nor passive; female primates are, on the whole, assertive sexual partners.

As Clementine passes into the main stages of adulthood, her life will include a few months of the year as a cycling female ready for mating in the fall. After a cycle or two, she'll most likely be pregnant. A female macaque gains at most a few pounds during a pregnancy, but Clem is small enough that her weight gain will show. Humans, who think they give birth to such large infants, see pregnancy as a time of great burden—the weight gain causes all sorts of physiological changes that make life a little slower. Other primates, who have lower-weight babies, seem to go about their lives in the usual way. In my groups of Barbary macaques, I remember only one female who seemed debilitated, the highest-ranking or alpha-female La Lune. In the spring of 1987 she was a whale of a monkey. She spent almost all of her day lying down, eating here and there. On April 5 she gave birth to healthy twins, a male and female christened Harold and Maude by us observers. It seemed that her lackadaisical last months of pregnancy had been warranted.

Compared with other mammals, primate mothers have it rather easy (Prentice and Prentice 1988). A female mouse, for example, produces several infants at a time during an extremely short gestation, and then she nurses them to large-mouse size during an abbreviated nursing period. The total stress of any pregnancy and lactation is actually much less for primates because of the slow-growing fetus and because the infant takes its time nursing for months or years. But still, the road to primate female reproductive success does have restrictive costs. Surprisingly, pregnancy isn't the problem—it takes only 125 extra calories every day to produce a human infant, for example. But lactation takes a larger toll—nursing mothers require twice their normal number of calories to feed an infant and keep it growing. Thus a woman who normally eats about 2,000 calories a day will use up 4,000 a day when lactating. The same is true for nonhuman primates: lactation is a draining time for females. Several studies have shown that nursing monkey mothers eat more and move less than their nonnursing contemporaries or that they lose weight during the period of lactation (Altmann 1980, Small 1982). And this need for hard-core nutrition cuts into their social time. They have fewer interactions during the day, and often pare down their close connections only to kin (Dunbar and Dunbar 1988).

Primate infants are dependent in other ways. They'll cling to the mother's fur and ride around, adding to her caloric expenditure. They need defending both from inquisitive troop mates, rivals who would just as soon remove them from food competition, and predators who see them as an appetizer. And mothers often have to intervene in play sessions that become too rough. Jeanne Altmann found that lactating baboon females more often died than nonmothers (1980)—making motherhood is a potentially dangerous life stage indeed.

The Golden Years

Clementine's life will continue at this pace for several years, a cycle of mating, pregnancy, motherhood, and sisterhood. She'll sleep in the trees each night near her relatives, move along the grass each day, and dig for shoots in the winter. Grooming her sisters and friends will occupy some of the day, but bouts of aggression and all-out fights will pepper her life with excitement and real danger. Her mother will grow old and die, and perhaps Clem will become the leader of her own matriline. She might even live past reproductive age and become one of the old arthritic matriarchs of the group.

We don't precisely know the life span of most primates. In captivity, from studies which give us the most information about the physiology of growth, animals live long lives because they are protected from diseases and predators. Old females also continue to reproduce, although their rate of reproduction sometimes slows down with age (Small 1984a, Watts 1991). In the wild, predators are likely to scoop up an aging animal as it becomes slower and more peripheral to the troop. The literature is replete with statements like "and so-and-so was never seen again. She is presumed dead." Stories of old females, far and few between, have some commonalties, however. Older females don't always lose power (although they do lose rank among Indian langurs—with them rank relates inversely to age); some continue to maintain the force of the matriline into old age. A macaque like Clem might at twenty-five or thirty years of age still be a member of her lineage. But there will be subtle differences in her behavior. Old females tend to peripheralize themselves from the rest of the troop, spend much of their time alone, and move more slowly. But these old females also have a use in the group. Peter Waser was the first primatologist to write about an old female, a gray-cheeked mangabey who was part of his study group in the Kibale Forest of Uganda. She was rather peripheral to the group, and her movements were slow, but she was treated with deference by most group members. She was also responsible for leading the group into new areas, or at least areas that were new to Waser, and she was the only adult female to initiate feeding on a resource that hadn't been eaten during his study (1978). Marc Hauser watched an old but very dominant female vervet monkey display a new way of eating (1988). During a time of very stressful ecological conditions, this matriarch used dried acacia pods like a sponge to soak up some gummy sap, a good nutrition resource, from the acacia tree. No one had seen this highly efficient behavior performed during the six years of observation on this troop. Hauser believes that the old female either invented the procedure or that she remembered it from times long ago. In either case, the spread of her method among her family and many other troop members made her a vital teacher during times of nutritional scarcity. Although there's no specific period of menopause in non-human primates, the aging process allows time for old females to aid their relatives because they have few infants left to care for. Even old langurs who have lost power because of their age are still among the first to help infants in trouble (Blaffer Hrdy 1981b). Clem may find herself carrying a granddaughter or defending a great-grandson. She'll be a functional member of her troop until the day she dies.

What's Important

The life cycle of a female macaque is punctuated by birth, sexual maturity, mating, pregnancy, social relationships, and death. Those features are interwoven in a complex fabric that results in a female's eventual reproductive success. It's almost impossible to calculate how any female will turn out in the end. Too many variables, too many chance events, and the irritating interference of others and their needs, prevent a female from having a smooth path to high reproductive success. Each female must navigate through life, and there are winners and losers in the game of reproductive success. But it's the navigation that's of interest to the observer. How do females travel through the complexities of life and end up passing on some genes to the next generation? What motivates some strategies over others? How do the strong survive?

The next several chapters focus on one state of the life cycle, mating. Mating is the moment of truth in reproductive success. A female who stays alive forever without mating isn't represented in future generations. For females, the decision to mate, and with whom, is a major key to successful reproduction. The next chapters try to explain exactly how female primates, including humans, use their available mating keys to unlock the gate to reproductive success.

CHAPTER 4

Female Choice:
The Theory of Mating

THE LARGE PINK REAR END of a female monkey bobbed through the juniper scrub. She was in estrus, or heat, and the swelling signaled to males, even those at a great distance, that she was ready to mate. She approached a young male, swung her rear into his face, and he mounted her. Both of them ignored the rattle of paper and scratchy pencil noises I made as I stood only five feet away, writing down their names, who started the interaction, and what happened once the sexual encounter ended. In 1986 I had come to this group of Barbary macaques, living on 20 acres of oak forest in southwestern France, to study female mating behavior. My job was to follow one of fourteen females for half an hour, several times a week, and note whom she mated with. The mechanics of this assignment were relatively easy because I learned early that Barbary females and males are clear in their choice of sexual partners. Barbary macaques breed only from September to January, when mating becomes the main activity of the adults. As females come into estrus, the skin on their hind ends inflates like a balloon in response to hormonal changes. These huge pink swellings indicate not only sexual receptivity but the fact that ovulation is imminent. A female Barbary selects a male by turning her hind end into his face, and if he's interested, they copulate. Some males wait for females to come calling, while other more assertive males approach females and give them a slight nudge from behind; if she's willing, they copulate. The female typically emits a loud, passionate call during the copulation, and then she spends a few minutes grooming him after it's over. The female always ends the interaction by moving on, usually making a beeline to the next male.

My framework for this study was a theory of female mating strategies based on traditional evolutionary principles. This theory, called sexual selection theory, has a long history that includes the notion that females are discriminating sexual partners. Recently evolutionary biologists have discarded the image of females as passive creatures and have embraced the notion that females, just like males, not only have been selected to attend to their own reproductive interests during mating but also that they are assertive sexual beings. But, theoreticians point out, females, especially mammalian females, also invest more heavily in offspring than males do; females gestate and lactate while males are free to inseminate other females. The biological fact of this difference in investment suggests that females should be relatively careful about selecting potential fathers. Therefore female choice of a mate should be critical to female reproductive success. It might also be a potentially significant evolutionary force for males, affecting the passage of particular male genes on to future generations. The immediate goal of my study on the Barbarys was to gather enough data to test hypotheses, or expectations, about female choice of mates. I knew that they mated with several males, but presumably they were more choosy when ovulation was imminent. I intended to observe the choices my females made and to discover why they preferred certain males over others.

Initially, the behavior I gathered on Barbary mating fitted nicely into this model of female choice. Clearly Barbary females decide who gets to mate and when; they are perfect examples of the sexually assertive female primate. But at the end of the breeding season, 506 copulations later, I found myself questioning the current party line about female choice. Yes, these females were making choices, but they seemed to choose every male in the group, one after another, and there was no selectivity during the time when ovulation might be occurring. And as the days in estrus increased for any female, either by long cycles or by a greater number of cycles, I found a corresponding increase in the number of different male partners she copulated with. If Barbary females are supposed to be selective about which males would father the next batch of infants, I asked myself, why are they moving from male to male with apparent indiscriminate abandon? My data supported only the part of female choice theory which indicated that females would be assertive sexual partners, but as for their choosiness—I was at a loss to explain why these females were more promiscuous than choosy.

I suspected that within the larger history of sexual selection and female choice theory were clues indicating why these females, and perhaps other primate females, didn't exactly fit current theoretical expectations. The day I watched a Barbary female copulate with three different males in the span of six minutes, I knew that it was time to reevaluate the current concept of female choice. Presumably the reason Barbarys didn't fit the standard model was because the model had some significant flaws. Thus began a historical search to understand why female choice theory said one thing and the females themselves did another.

Sexual Selection Theory

The theory of female mating strategies begins, as do all frameworks for animal behavior, with Charles Darwin. When Darwin published *The Origin of Species* in 1859, he sparked a revolution in the way we look at animal species, including humans. Species evolve, he suggested, because some individuals are more successful at staying alive and producing infants than are others. Traits and behaviors that aid an individual in survival and reproduction are selected over generations and mold the morphology and behavior of a species. But while promoting his theory of evolution by natural selection, Darwin was aware of a major flaw in his argument. If all individuals were selected by the same forces in a particular environment, why did males and female of the same species look so different?

He could account for some differences in male and female characteristics, such as ovaries, testes, sperm, and eggs, because those differences are necessary for conception and pregnancy to occur. They're what Darwin called "primary sex organs." If a species reproduces sexually, by definition there must be differences in the reproductive organs. Different primary sex organs have been selected by natural selection over generations to accommodate the complex process of reproduction by two cooperating parents. But more difficult to explain was the appearance of exaggerated features that pop up in only one sex and are often a hindrance to the individual. They appear at puberty yet have nothing to do with conception. And sometimes even when both sexes have the trait, it's exaggerated only in one sex or the other.

The peacock's tail is used most often as an example of an exaggerated characteristic found in only one sex. If you walk into the Burggarten in Vienna, Austria, a baby's plaintive cry often greets you near the statue of Mozart. You wonder if someone's neglecting a child, or if a baby needs a change. Strutting in front of the statue is a male peacock, and it's his cry that sounds like a wailing child. If disturbed, the male will stare you down, and then explode his tail into a dazzling fan of feathers like a Las Vegas showgirl. Walking nearby, probably unnoticed by you, is a possible mate of the male—the drab peahen. She's dull brown, she has no flashy tail, she doesn't strut or preen. She looks almost like a different species of bird. Or think about male and female lions. The females are smaller and they lack the regal brown mane of males. The male greater kudu of East Africa sports long twisted horns that reach two feet into the sky, while the female looks more like Bambi. If natural selection is operating on all individuals of a given environment equally, why should these males have things that females don't? Or put another way, why are males burdened with these extra accoutrements?

Darwin's answer to this question was simple. He suggested that males and females often look different, and sometimes behave differently, because they operate under different sexual pressures. Males must often fight one another to gain access to females, and, he proposed, because females are less interested in sex, males must spend time and energy attracting reluctant sexual partners. The exaggerated physical feature of males and their sometimes seemingly demented patterns of behaviors, like courtship dances, function primarily during the mating process. According to Darwin, these traits and behaviors have been "sexually selected" rather than naturally selected (1859, 1871). One can consider a trait sexually selected if it appears only in one sex, is used only during reproduction and aids the individual in gaining mates. This selection has nothing to do with the struggle for existence but everything to do with the struggle for passing on genes. Under the process of sexual selection some individuals gain the best, or the most, mates while others are less fortunate. Those who do best are the winners in the game of reproductive success, and their genes are passed on in higher frequency. And the traits or behaviors that helped them be successful competitors are passed on more frequently than are the traits of the less able individuals. According to sexual selection theory, the "extras" seen on males, for example, aren't extras at all. They are weapons, armor, and ornaments that help in the battle of passing genes on to the next generation, and their sexual selection over time aids those winning males.

For Males Only

Most of what Darwin proposed as sexual selection makes immediate sense. If traits or behaviors arise only after males reach puberty, and if those traits help gain mates, those attributes will be passed on as successful males conceive more offspring than do males with smaller versions of the same trait. But how exactly does owning some sort of special equipment help a male?

Darwin offered two channels by which such traits could be passed on (1859, 1871). The first way, he proposed, is through competition within each sex, or what's now called intrasexual selection. The antlers of male deer are a case in point. When the rutting season begins for white-tailed deer in New York State, the woods often echo with the crack of antlers smacking together as two males crash head on. They are fighting for access to females who are in heat for only a short period. One male postures better than the other, one is able to push the other farther, and pretty soon the losing male backs off into the woods. The winning male then copulates with nearby females, distributing his genes, including the ones for good-sized antlers and strong fighting ability. The other male wanders through the forest, looking for another combatant with whom he'll have a better chance at winning. Darwin noted that competition within one sex for access to another sex would drive the exaggeration of traits that might help in competition, and he observed that males were most often the combative sex.

Examples of the intensity of male-male competition, and the real results of ousting the competition and gaining more mates, abound in nature. Massive horns in male bighorn sheep are used to push other males into reproductive oblivion; large canines are slashed about by a male baboon trying to keep other eager males away from his consort partner; bulbous throat pads wielded by male elephant seals protect them from the slam-dance of other males during their male-male battles on the beach. In all cases, males compete with one another for females, while the females stand by waiting for the males to end the fighting and begin copulating. Such battles are especially fierce in arenas where females cluster together and come into heat simultaneously, creating a raucous male circus revolving around fertile females.

Nobody seems to ask much anymore why males have these battles: the evolutionary explanation is so clear. The battles occur too often, and the consequences of the skirmish is too predictable, to cause much theoretical bickering. Animal behaviorists have, in fact, proved one of Darwin beliefs. Many exaggerated traits seen only in males are used solely for mating and have been driven to exaggeration by sexual selection in the form of male-male competition.

For Females Only

The other channel suggested by Darwin that might explain why males and females look and behave differently is less clear. It's also the one most misconstrued by behaviorists and evolutionists—mate choice. Here Darwin was on shaky ground, and the past 120 years haven't improved on Darwin's initial thoughts very much. Darwin suggested that selection for some odd male traits, and thus the differences between males and females, could be explained by competition by one sex for the attentions of the other sex. This is now often referred to as intersexual selection, or mate choice. Most biologists agree that when it comes to choosing mates, females and their preferences will be more important in the evolutionary sense than the mate choices of males (more on this later). Thus intersexual selection is sometimes referred to specifically as female choice. In this scenario, females can be differentially attracted to some males over others, prefer them for some reason, and make an active choice for a certain type of male. Males with brightly colored features, pretty songs, or silly dances, for example, might win the favors of females simply because they catch the eye. There's no battle here, just attraction and choice.

Darwin based his ideas on female choice on the assumption that females are passive, less eager sexual partners. This supposed lack of passion on the part of females, he believed, explains why secondary sexual characteristics appear more often in males than in females. Exaggerated and unusual traits are needed only by the sexually potent male sex because they alone were motivated to fight for, or attract, the opposite sex. Because males more often have secondary sexual characteristics, he reasoned, male-male competition in general must be a more powerful selective force than is female choice.

In his writings in *The Descent of Man and Selection in Relation to Sex*, published in 1871 to explain human evolution, human racial variation, and sexually selected traits, Darwin was very clear that female choice has only a minor role in explaining odd male traits. He devoted few pages to female choice, and bluntly called it not important. He also believed that female choice occurred only *after* males had battled it out. And even more depressing for females, he questioned their brain power to make decisions at all: "Hence the females, supposing that their mental capacity sufficed for the exertion of choice, could select one out of several males" (p. 259). He thought that females were rather passive about the whole thing: "Or she may accept, as appearances would sometimes lead us to believe, not the male which is the most attractive to her, but the one which is the least distasteful" (p. 273). Poor females are caught in this biological bind of having to choose a mate, then actually go through with a copulation. Female choice, for Darwin, evolved only because males need to attract females. Once again, he used sexual selection, this time female choice, to explain the appearance of male traits and behaviors. Ones that didn't evolve by male-male competition, he suggested, must have appeared as seductive attractors.

No one, including Darwin, was interested in females themselves. After all, scientists figured, fertile females can always conceive, and because they are apparently less eager for sex, females will always be in demand. Darwin even went so far as to suggest that females don't usually have any interesting traits to explain away because they just aren't endowed with the passions of males. According to Darwin, males are the eager, passionate gender, and thus only males evolved traits and behaviors to help them out sexually.

Sexless Victorian Females

This view of females, as passive and sexless, was accepted by Darwin and held by the scientific community for a century. Why was it so easy for scientists to accept a picture of females as asexual, passive individuals tossed about by the throws of sexual selection? The social milieu, both then and now, has had an inordinate influence on how scientists view female behavior.

Like all scientists, Darwin was influenced by the Victorian world in which he lived—a world in which women of good breeding were dependent, passive, and supposedly asexual. Perhaps this background accounts for the following words he wrote about female sexual behavior: "The female, on the other hand, with the rarest exception, is less eager than the male. As the illustrious Hunter long ago observed, she generally 'requires to be courted'; she is coy, and may often be seen endeavoring for a long time to escape from the male" (Darwin, 1871, p. 273). There were not many accounts of female mating behavior in those days—most observations were anecdotal or concerned domesticated animals that copulate only when allowed to, or forced to. No one really knew the "truth" of female mating behavior. The Victorians believed that only prostitutes, motivated by money, and nymphomaniacs, who were pathologically driven, eagerly engaged in sexual activity. What women Victorians felt, or what they did behind closed doors, is mostly a mystery (but see Brown Blackwell 1875, Burt Gamble 1894). Thus we can't really blame Darwin for his perception of females—it was one shared by most of the English gentleman of his day, reinforced by the position of women in Victorian times as intellectually inferior but morally superior individuals whose role was to keep the household functioning (Sleeth Mosedale 1978).

Darwin was writing about the "nature" of female mating behavior without any concept of female sexuality. His views of animal behavior reflect what he saw at home, at least in a general sense. As Evelleen Richards, a feminist historian, points out, he was surrounded by a perfect Victorian family (1983). His wife, Emma Wedgewood Darwin, an intelligent and capable woman, bowed both to his scientific career and his illnesses. The women in Charles's life were treated as inferior—his daughters were barely educated and yet the sons were well educated. Although Emma disagreed with his evolutionary views and worried about his lack of belief in God, this difference of perspective never got in the way of their relationship (Litchfield 1915). Interestingly enough, Emma was never converted to his views, although she understood his work (Richards 1983). And yet it was to Emma that Charles entrusted his draft of *The Origin of Species* in case he didn't live to publish it. Emma maintained a household built around his health and wishes and he apparently worshiped her for that care.

Thus it seems reasonable to suggest that when Darwin wrote, "Man is more courageous, pugnacious, and energetic than woman, and has a more inventive genius," and, "Woman seems to differ from man in mental disposition, chiefly in her greater tenderness and less selfishness" (1871, pp. 316, 326), he was presumably echoing his own life and relationship with women. There was no contradiction in what he proposed for the natural world of female animals and what he saw in his immediate vicinity. Females were, by nature, coy, choosy, and very reluctant sexual partners.

Today we should know better, but the long arms of Victorian sexism reach into contemporary biology, especially in its thinking about sexual selection, female choice, and female sexual behavior (Hubbard 1979, Richards 1983). Although females are now seen as active sexual participants, there's always a footnote to their sexuality that makes females somehow "less sexual" than males. Modern texts in evolutionary biology and behavior concede that females may be sexually motivated, but a female's "need" to be choosy and not waste a precious conception renders her almost impotent. The issue now centers not on the rate of sexual interaction or the degree of sexual motivation by either sex but on the number of different partners. According to modern evolutionary theory, males "should" still copulate as often as they can and with as many partners as they can sequester, but females "should" be careful, choosy, less eager to mate. This view is, perhaps, only a modern, post-sexual-revolution version of Victorian sexism.

Female Choice—On the Edge of Biology

There's no need to take offense at Darwin's misogynist view—no one at that time really paid attention anyway. Alfred Russell Wallace, fellow evolutionist and well-known naturalist of the day, believed that Darwin was completely out of line for even suggesting female choice had any influence on evolution at all. The rest of the scientific community agreed with Darwin at least in principle: females had little influence on the mating game. And what Darwinian theory predicted was apparently seen in barnyards and forests: most male animals fought with one another, and the female reluctantly picked out a winning male to be the father of her latest brood. All thought of sexual selection, including mate choice, was therefore of little concern to biologists for the next sixty years after Darwin's original formulation. It's impossible to figure out why some ideas gain prominence and attention and others don't. It's possible that evolutionists were so concerned with explaining and defending the theory of natural selection to the world that sexual selection, let alone female choice, was not the major focus for biologists (Maynard Smith 1991).

During the 1930s and 1940s, when geneticists drew up the synthetic theory of evolution and Darwin's evolutionary theory was finally combined with modern genetic theory of the modes of inheritance through chromosomes and genes, sexual selection and mate choice came up for a temporary breath of air. A mathematician, Ronald Fisher, focused on female choice for a few pages in his book *The Genetical Theory of Natural Selection,* published in 1930, but like Darwin, he was interested in female choice only insofar as it influenced the evolution of male traits. He began with a little hand slapping. Fisher maintained that naturalists, who were still not particularly interested in sexual selection theory anyway, had become so focused on differentiating male-male competition from female choice that they'd lost sight of the unifying principle underlying sexual selection—competition for mates (1930). The distinction between the two types of sexual selection, male-male competition and female choice, had become important, he said, only because of the thorn of female choice. Wallace had dismissed it, and nobody was particularly convinced that females might have an effect on the evolution of male characteristics at all. But Fisher's major contribution to female choice theory, as we know it today, was the insistence that certain requirements must be met if mate choice, particularly female choice, is to have an evolutionary effect. First, the preference must occur consistently. It doesn't really matter if one female chooses this and another chooses that: females in general must prefer one type of male for their choices to have an evolutionary effect. Second, the preference must confer a reproductive advantage to the chosen individuals. It's not enough to demonstrate that most females reach out to some males with a particular trait for that trait to be favored by sexual selection; the said males, favored by most of the females, must also have lots of offspring. He saw the process as a kind of coevolution—males have a trait, females like it and mate with them. Females then have sons with the trait and daughters with the preference. The trait and the preferences are inherited together over and over, coevolving with each other (O'Donald 1980).

Fisher also reiterated another point made earlier by Darwin: there must be checks and balances in the sexual selection system. If females are drawn again and again to males with an outlandish trait, the trait would eventually appear larger and crazier after many generations of favoring only those at the outer limits of exaggeration. The reason that sexually selected male traits haven't gone wild is that natural selection waits in the wings, ready to end the path of runaway selection and keep things in moderation. For example, males might evolve giant canines that aid in their battles over females, but eventually these big teeth would get in the way of normal eating, and regardless of the advantage of large canines to male-male competition, the limits of natural selection would be reached. Thus natural selection is the brake that keeps the runaway train of sexual selection from going off the evolutionary cliff.

The direct relationship between female choice and the exaggeration of a male trait has been empirically demonstrated for African widowbirds by Malte Andersson (1982). Male widowbirds are black and have red epaulettes. The females, by contrast, are dull brown. The males also sport an extremely long tail, about 50 centimeters, while the females have a more normal-sized appendage. In an ingenious experimental design, Andersson added pieces of tail feathers to some males and shortened the tails of others; for controls, he left some males as they were or clipped and then glued back the same tail segment on the same bird. Andersson found more new nests in territories owned by males with artificially lengthened tails. He also shows that males didn't use their exaggerated tails in male-male competitions, and he concludes that the only explanation for the evolution of long tails is female choice. Female widowbirds consistently choose males with the longest tails, thus demonstrating the power of attraction of long tails to the opposite sex, although the elongated tail in and of itself has no real value to males or females and might even hinder the male.

Fisher's work might have led to a renaissance for female choice theory, but still the times weren't ripe. No one had yet conducted a long-term study of female behavior, and there were few studies of mating behavior in particular. But Julian Huxley, the zoologist, took up the gauntlet in 1938. In two important papers Huxley wrote a thorough evaluation of Darwin's original sexual selection theory (1938a, 1938b). Like any good scientist he questioned Darwin's assumptions. Huxley proposed that most male characteristics, be they aggressive displays or morphological features aimed at competitors, could be explained as adaptations that "promote the union of gametes." And like a good scientist he also tried to give this process another name—epigamic selection—to underscore the need for these features in reproduction. No one uses this term today because it's superfluous. He really meant selection for features needed for male-male competition, just as Darwin did. While Huxley was grinding away on the details of male-male competition, he seemed to dismiss female choice. Huxley, too, believed that mate choice, and this meant female choice, was a minor evolutionary force. He explained most ritual courtship displays as events in which males rev up their sexual engines rather than try to win the favors of females. Although these papers are considered important for the history of sexual selection theory, they are, as the geneticist Peter O'Donald says, "hopelessly confused" (1982). More to the point, these papers added nothing to the concept of female choice in particular and actually sidetracked researchers even farther away from female mating behavior. It would be almost twenty years before an insightful biologist began to wonder seriously about the female role in mating.

The Unheard Champion of Female Choice

In the 1950s, John Maynard Smith, a biologist at the University of Sussex, began working with *Drosophila*, the tiny flies that invade your rotting fruit in the summertime. Although Maynard Smith was initially interested in male fertility and the mating process in general, his early lab experiments on flies resulted in some unexpected information about female mating behavior.

In one of his first experiments, Maynard Smith worked with an inbred strain of flies in which many of the males became infertile because of their inbred heritage; the overall population of male fruit flies varied in relation to the degree to which each could fertilize eggs (Hollingsworth and Maynard Smith 1955). The variation in potential male fertility also fortutitously led Maynard Smith down the female choice path. During the experiments, he noticed the ritualized courtship dance of male and female flies. Females and males face each other and dance right and left in a coordinated fly tango. At first, when females bobbed right and left in front of a courting male, Maynard Smith assumed that the females were trying to get away. But experiments with wax models of female flies in another laboratory showed that female participation in the dance was actually required for mating to proceed smoothly. More important, Maynard Smith noticed that the female often started dancing first, just to see if the male could keep up. When faced with inbred males, who were clumsy dancers with bad timing, females turned away. These inadequate suitors sometimes tried to mount females, but they were often too far forward or too far back, and the female would eventually kick them off altogether. As Maynard Smith put it for these poor males, "The spirit is willing but the flesh is weak" (1955, p. 272).

He proposed that the female dance, which must be closely followed by the male partner if he wants to achieve copulatory success, evolved on the female's behalf. In other words, she judges a male by his dance, and the dance is a true indicator of male fertility. Presaging the words of biologists who were still in grammar school, Maynard Smith predicted that females might have interests of their own on which they base mate choices. Second, these female choices might have an effect on the appearance of male characteristics and behaviors. This was a new kind of female choice, one not centered on the importance of male-male competition or on the notion that females mindlessly choose males with the wildest displays. This was not Darwinian or Fisherian female choice in any sense because it viewed choice from the female side. Maynard Smith was suggesting that such choice also had a major impact on the evolution of the species.

Unfortunately, no one was listening. Maynard Smith recently wrote: "When, in 1956, I published a paper showing at least to my own satisfaction, how female fruit flies choose males, I do not remember receiving a single reprint request" (1991, p. ix).

It would be twenty more years and the one-hundredth anniversary of Darwin's concept of sexual selection theory before anyone seriously considered females, and female choice, as a powerful selective force.

The Feminist Revolution in Animal Behavior

The average person on the street is not aware of revolutions in academics. No one is killed, no one dies, although many a reputation can be wounded. In the late 1960s and early 1970s, a major change was in the wind for behavioral biology. Studies of animal behavior before that time had been thought of as a kind of natural history—watch animals, describe what they do, paint a picture of the animal's life. But as I mentioned in Chapter 3, several biologists during the 1960s had begun to look at evolutionary explanations for behavior. Eventually the discipline of sociobiology, the study of the biological basis of social behavior, was born. The tenants of sociobiology are simple. Patterns of behavior, just like skeletons and muscles, have evolved by the rules of natural selection and sexual selection. Individuals who behave in ways that improve their reproductive success pass on more genes than do those who behave in ways that don't help them pass on genes.

Females, perhaps, benefited most from this new perspective. Before the days of sociobiology, female animals were considered only as mothers, and no one really thoroughly investigated what else females did with their lives. The new evolutionary framework allowed for hypothesis building, and females were part of the equation that would eventually explain patterns of behavior. This reorientation toward females also caused sexual selection theory, and female choice theory, to gain attention.

Interestingly enough, this flowering in the history of female choice theory isn't just a shift in an established scientific paradigm. It can also be accorded to major changes in the social milieu of the very biologists who were developing theories of animal behavior. In the 1960s and 1970s, two other revolutions were in progress in Western culture, the feminist revolution and the sexual revolution. Although there's no way to prove that an orientation toward nonhuman female behavior and sexuality was initiated by these not-so-subtle cultural forces, no one can deny that the new social freedoms for women in the Western world had some sort of influence on the new chic position of female animals in behavioral studies (Andersson and Bradbury

1987). In addition, more women were entering the sciences, and although not all women behaviorists studied females, and many men did, there was an air of refocus.

But don't be deceived. Academic feminists of the time did not sanction this reorientation and they did not clearly understand (nor do they now) biology's role in explaining female behavior. The anthropologist Sarah Blaffer Hrdy wrote about her graduate education at Harvard in the 1970s and the odd position of being a feminist studying biology: "Within the Harvard of that time there was no overlap at all between feminism and evolutionary biology, not even a common language. Feminists were outraged at what they took the sociobiologists to be saying, and the sociobiologists were mystified to discover that feminist were demonstrating at their lectures. As a woman in the midst of all of this, I felt torn and often quite alone" (1981a, p. ix). Studies of female animals were gaining ground, but many things about the "natural" behavior of females disturbed their human cousins. For example, field and lab studies showed that male animals were often highly aggressive and dominant to females. But what the feminists were slow to hear was the female power inherent in many behavioral descriptions. Blaffer Hrdy's book *The Woman That Never Evolved* was revolutionary for its time because she said that female primates were not passive individuals waiting for males to finish battles. Instead, she painted a picture of competitive, strategizing, sexually assertive creatures (1981a). According to the new evolutionary biology of those years, female primates are the same as males in that they have been selected to act in ways that improve their reproductive success. At the same time, sociobiology showed that females are also different from males because the constraints of pregnancy and parenthood have molded very different creatures. Females and males are apples and oranges thrown into the same basket.

And in the midst of the new perspective on behavior, where was the issue of female choice as a potential evolutionary force? Gaining ground.

The major boost to female choice theory had occurred in 1972, when Bernard Campbell proposed an edited volume to celebrate the one hundred-year anniversary of the publication of Darwin's *Descent of*

Man. Although many papers in this book are important and interesting, none has become more famous in biology circles than Robert Trivers's work on parental investment and sexual selection (1972). Trivers followed John Maynard Smith's lead of two decades earlier when he turned his attention toward female choice. But he also added a spin to female choice which may have had serious, and I think somewhat negative, effects on the way biologists view female behavior.

Two Kinds of Choice

Trivers explained something Darwin couldn't. When Darwin first thought of sexual selection and female choice, he reasoned that the passionate males would be the fighters and the less eager females the choosers. Trivers, with one hundred years of biology under his belt, was able to support Darwin's dichotomy on a purely biological basis without making reference to passion. Trivers pointed out that males produce mass quantities of highly mobile gametes, called sperm, and because these cells are available in the millions, males have the ability to inseminate many females, so that each male's potential for passing on genes is extremely high (1972). Females, on the other hand, normally produce few gametes, even if they have litters or broods, and those eggs are precious reproductive commodities for females. This division makes evolutionary sense and can easily be applied to Darwin's original division of the sexes in terms of sexual selection; males will compete and mate indiscriminately and females will be choosy because of their differences in reproductive potential. But Trivers went further. He pointed out that beyond the gamete level there are significant reproductive differences in males and females because the two sexes differ in the amount of parental care they invest in offspring. Each individual has only so much energy to expend on bringing up offspring. For most females, especially mammals, reproduction includes not only the production of fewer gametes but also gestation, nursing infants, caring for them, and protecting offspring from predators. A female's lot, or better, her evolutionarily selected strategy, is such that she must invest heavily in infants. Males, on the other hand, can go about spreading sperm as long as there are females to mate with and infants don't need any fathering. Females are the limiting resource in this mutual scenario because, after insemination, females must drop out of the mating game and attend offspring. Males, on the other hand, will always be fighting over females because males need females not only to bear but to bring up infants. Trivers didn't address the origins of this reproductive difference, he just pointed out the consequences—females should be extremely choosy about whom they share their genes with.

Darwin and his colleagues originally believed female choice to be a simple matter: females look out for the right species, the right sex, and the right age male—perhaps waiting for the battle winner or choosing a mate with an odd trait. Darwin certainly never dreamed of the kinds of discriminations Trivers proposed. Female choice, from 1972 on, imbued females with the power to choose the "best" males on every level—genetically, physiologically, and behaviorally.

Trivers's work is important because it veers away from the established view of female choice. Previously, female choice had been used to explain odd male characteristics; Darwin, Huxley, and Fisher never considered the possibility that females might have interests of their own. After all, they thought, an estrous female can always conceive because there are zillions of sperm lurking in every corner. Trivers, surely influenced by his own social milieu, pointed the spotlight on females themselves. Under Trivers's scheme, a female's choice might also drive the evolution of odd male characteristics, but it might also be a choice made by the female in her own best reproductive interest.

Contemporary Female Choice Theory

This simple concept, now called "female choice," has riveted researchers in animal behavior since the early 1980s. Behaviorists find it difficult to demonstrate, and yet we intuitively know that females somehow influence who gets to mate and when. We have usurped Darwin's original meaning, added a few twists of our own, and ended up completely confused by the whole issue. In essence, we have made "female choice," as an evolutionary concept, as illusive as female sexual behavior was to the Victorians.

The result of Trivers's theorizing, in particular, is that we have two ways of looking at female choice (Heisler et al. 1987), although these are usually lumped into the general heading "female choice," and most biologist often don't see, or distinguish, the difference between the two types. In the first kind of choice, proposed by Darwin and refined by Fisher, females make choices, for whatever reason, and after many generations the male trait and the female preference co-evolve until the trait is exaggerated and useless. It doesn't really matter why the female chooses a particular male, only that she and her female colleagues keep on doing it. Eventually the male trait no longer has any fitness value to either partner but males keep up the waving and females continue to choose. The result is the exaggeration of a trait that means nothing in terms of reproductive value or as a sign of male vigor. I call this way of looking at female choice "Fisherian" choice.

Some biologists see this approach to female choice theory as "nonadaptive" (Kirkpatrick 1987), because such choice can lead to the development of useless traits or to no development at all. This point that the female action does not necessarily influence the male trait per se is powerfully illustrated by Andrea Basolo's work on green swordtail fish (1990, 1991). Given a choice, female swordtails consistently choose males with longer tails, and Basolo shows that only the length of the swords, and no other perceivable male traits, has any effect on female preference. But more important, she shows that females in ancestral swordless platyfish also prefer sworded males when conspecific males appear with artificially attached swords. In spite of this inherent preference among platyfish females, however, there's been no evolutionary change in the morphology of caudal fins in male platyfish. Basolo's research establishes the existence of a female mating preference that has had no apparent evolutionary effect on male phenotype. In such cases, "female choice" isn't synonymous with intersexual selection. The demonstration of a historical evolutionary *enhancement* of a male character, not the existence of the preference, is the first stimulus for considering sexual selection as the mechanism of change. Mathematical models have used this type of female choice to test how male traits might evolve (Boak 1986, Kirkpatrick 1982, Lande 1981, O'Donald 1980, Seger 1985), but models don't explain what females and males actually do, or how they pair off and mate.

The other way of looking at female choice, by which females have their own interests in mind, and the type of choice most biologists now attend to, is what I call "Triverian" choice: females make choices based on what they perceive as best for them, choices that may not have any effect on males. The only time a male is affected, as in the Fisherian sense, is when those traits somehow have a link to the fitness of a male (Small and Palombit ms., Small 1992b).

For an example of the subtle difference between Fisherian and Triverian female choice, we turn to pheasants. Torbjörn van Schantz and colleagues from the University of Lund, Sweden, have shown a direct connection between a male trait—the spurs on the legs of male pheasants—female choice, and improved reproductive success of females (1989). They factored out male size, wing span, territory, and age by changing the kinds of males available to females. Only spur length was of consistent importance to females. And females who mated with long-spurred males hatched the largest clutches. In other words, if a male trait is correlated with health and vigor, and females are attracted to that trait, their choices will affect the reproductive success of the females and the evolution of the male trait. In this case, the genes for male spur length are passed on to her male offspring, but this fact is almost incidental to the fact that females who choose these males have bigger clutches. These traits are called "truth in advertising" on the part of males. But from the perspective of the female, the effect of a choice on male traits is not important. She is driven only by the result of that choice for her offspring. From this perspective, the male trait, and how it evolved, is secondary to the motivation of the female making the choices.

The proponents of the "truth-in-advertising" school of thought, also known as "the good genes" hypothesis (Kirkpatrick 1987), look at female choice from the purely female point of view, that is, the way females might look at potential mates. They believe that any characteristic a male waves purposely at females is really an indicator of male vigor (Andersson 1986, Hamilton and Zuk 1982, Kodric-Brown and Brown 1984). For example, a male deer with the biggest antlers is probably the healthiest and will pass on the best genes. Or the bright coloration of male birds can be explained as a neon sign selected by female choice because colorful males have fewer parasites and females are attracted to their brightness and good skin condition (Hamilton and Zuk 1982). This perspective suggests that females make fine-grained discriminations among males and choose those who potentially have the best genes. But males can play this game too. One evolutionary biologist, Anton Zahavi, has suggested that males sport these odd characteristics only as a metamessage to females (Zahavi 1975). According to Zahavi, a male dragging a huge tail is shouting, "Hey, honey, look at me. I can walk around *and* carry this unbelievable thing. Mate with me." The only truth in his advertisement is that the male can operate with a giant handicap.

Both these two types of female choices, or two ways to analyze why females make certain choices, have the potential of affecting male characteristics, but the "Triverians" don't really care. For them, if a female makes a choice of any kind, regardless of how it affects the evolution of male traits, it's female choice. And many biologists have even forgotten the connection between female choice and the possible evolution of male traits. They focus solely on female choice for improvement of female reproductive success. Patterns of behavior or attributes of males that might help this reproductive success, not exaggerated male traits are the focus of many studies today.

Some male traits may act as neon signs to females, and the appearance of these traits has certainly been affected by female preference over time. But both theory and real studies have made the issue of choice even more confusing. Many studies have shown that females do choose males for certain traits and presumably drive the evolution of those characteristics. To name but a few, female preference and choice can explain the loud courtship call of túngara and cricket frogs (Ryan 1983, 1990), male coloration in ladybirds (Majerus et al. 1986), or the pattern of orange spots on male guppys (Farr 1977, Houde and Endler 1990). But a direct connection between female choice and improved female reproductive success in the Triverian sense has been shown in only a small number of species such as pheasants and *Drosophila* (Partridge 1980, von Schantz et al. 1989). For most species, the impact of any kind of female choice is still so much theorizing. And certainly no one has clearly demonstrated how female choice might influence the evolution of a particular pattern of *behavior*. Can female choice select for the pattern of behavior called male "friendship," or bring about the evolution of paternal care?

We're actually unsure what females want, how they choose, and if they do make choices, how those choices affect individual female or male reproductive success. So far, the theory behind female choice has been basically a road map used to initiate studies. More interesting information, and intellectual growth, has come from what animals teach us with their behavior beyond the theoretical predictions. The original theory of female choice, and its development, has been merely a steppingstone to understanding what female animals really do when they mate.

The Art of Choice

The verb "choose" is defined in Webster's dictionary as "to take as a choice; pick out by preference from all available, select." When the word "choice" is used to describe human behavior, we assume it always includes some sort of decision-making process. A person is presented with a number of options, evaluates those alternatives, and then makes the selection. We Westerners like to believe that all our choices are conscious; we firmly believe in the concept of free will, and the idea that some decisions in life are unconscious often implies a lack of freedom. And worse, we'd hate to think some things were "fated" and predictable, and thus also out of our control! But humans, like all animals, make both conscious and unconscious choices, and some choices have a foot on each level of consciousness. For example, it's reasonable to suggest that choice for a certain sandwich at lunch is a conscious choice, whereas choice for a particular mate is probably more unconscious.

But when the word "choice" is applied to animal behavior, we're at a loss to describe exactly what the animals are doing. The animals can't explain their motivation in words, and we observers can rely only on their behavior as a clue to their preferences. In most studies of female mate strategies authors provide a qualifying statement to the effect that their subjects are not making "real decisions," but that natural selection has operated to favor those females who prefer mates with certain qualities. The action then is presumably unconscious. Only the primatologist Robin Dunbar, in his volume on gelada baboon female behavior, is straightforward in his opinion that the animals are making conscious decisions about their social interactions: "I shall make frequent use of the language of conscious decision-making. . . .

I do so partly because this is much the easiest way to discuss the animals' behavior, but also partly because fifteen years of field work have made it abundantly clear to me that strategy evaluation is precisely what the animals are doing" (1984, p. 4). It's probably more useful to avoid the semantic machinations involved in deciding whether mate choice is conscious or unconscious. After all, we'll probably never know how animals think or how they make decisions, and evolution doesn't care anyway. Evolution acts only on the consequences of the decision.

We get into another semantic puzzle when authors use the word "preference." While "choice" indicates an action that can be observed and measured, "preference" denotes desires that may or may not result in choices (Heisler et al. 1987). For example, a female might prefer the most dominant male, but she makes a choice for a less dominant one because she's afraid of the bigger male; she may prefer a particular male (desire or motivation) but be unable to gain access to that male (the choice itself). So far we know only that female primates do make choices, be they conscious or not. The whys and wherefores of those choices, and the preferences they're based on, are more difficult to figure out.

Unrequited Needs

Probably the major reason female choice theory has developed along such a twisted path is the nature of the behavior. A female's selection of a mate isn't a simple matter of evaluating a number of possible options, deciding on one, and marching up and taking him. Any female choice of a mate is complicated by the fact that she's interacting with males, and those males have agendas of their own. Not for one moment should we forget that female mate choice operates in tandem with male mate choice. Unfortunately for females, and those who observe nonhuman primates, the presence of males can have a domineering effect on potential female choice. With humans, we can at least ask a hypothetical question about what a woman might want from a mate if she lived in a free-for-all female-choice world. Asking nonhuman primates is more difficult; we must look for subtle clues.

Barbara Smuts, an anthropologist, has recently suggested that much of primate mating, including human mating, occurs in an arena of male sexual coercion laced with violence, and thus coercion may be true for other animals as well (1992, Smuts and Smuts in press). She illustrates her point with chimpanzees. Although the majority of chimpanzee matings occur opportunistically, females often follow males on safaris away from the group, and a third of all conceptions occur during these exclusive matings. Males try to tempt females away, but sometimes they refuse to go; the ardent male may threaten the female and attack her (Goodall 1986, Tutin 1979). Females usually respond to these attacks by following the male in question. Smuts's point is that these scenes may indicate long-term tension between males and females which place females at risk from coercive males. It may be that even when females seem to wander off under their own free will, they are responding to previous attacks and are following only out of fear. Attacks by males on females in many species increase during breeding seasons, and males routinely threaten females in estrus. Male macaques, for example, routinely interrupt consortships between estrous females and males lower in rank (Hanby, Robertson, and Phoenix 1971, Huffman 1987, Manson 1991), although females usually resume the consortship posthaste.

In one study, the possibility of pure choice by females without male interference was tested in the laboratory. Pigtail macaque females were trained to hit a switch when they wanted to release a male into their presence (Eaton 1973). Females released males most of the time, but seemingly for purely sexual reasons: the females copulated throughout their cycles with any male they set free. They were also rather easy with their favors—most females released three of the seven possible males they were paired with. And when one male was repeatedly aggressive toward the subject females, they just stopped releasing him. The point is, it's hard to tell the motivation of a female, and her preferences or choices, when there are males involved who are often larger and stronger and tend to bully their way into matings. In these situations, according to Smuts, free choice for females may be an illusion.

Females may also be frustrated by their own kind. If we assume that each female is out to improve her individual reproductive success, it follows that competition among females may occur, especially if preferred males are in limited supply. In seasonally breeding groups, there may not be enough males to go around, and it's possible that females mating at a high rate will deplete the sperm supplies for other females (Small 1988). Two studies have given preliminary evidence that this is true. A comparison across eighteen field studies of baboons shows that when the ratio of males to females decreases, the birth rate declines (Dunbar and Sharman 1983). Similarly, in a captive group of bonnet macaques, female fertility was lowest in the years when the ratio of females to males was the highest (Silk 1988).

Because the prevailing sentiment concerning male sexual behavior is that males will copulate at the drop of a hat and that any estrous female can conceive, why are some females not conceiving? Non-human primates show us what many single women in America today already know—sometimes it's very difficult to get a date. Female rhesus monkeys and baboons often present to males, a clear sign of preference and choice, but males regularly refuse (Lindburg 1983, Saayman 1970, Scott 1984). Lion-tail macaque females, especially sub-adults, share this rejection. Females of this species initiate almost 70 percent of the copulations but only 59 percent end up in mounts (Kumar and Kurup 1985b). No one is sure why these males refuse, inasmuch as sperm is supposed to be so cheap, but males often ignore estrous females. Thus female choice is frustrated by male whims.

Are the females really free to choose? Are they coerced by males and shoved aside by other females into relationships that aren't particularly favorable? Once again we must ask, when females have preferences, are they allowed to make choices?

Given Their Druthers, What "Should" Females Want?

The current consensus about the "shoulds" of female mate choice is very clear—females bear few offspring in which to pass on genes, and each is dependent for a long time; thus they should be careful when choosing fathers for those infants. Female choice for her own reproductive well being, when it occurs, should reflect the reproductive interests of females.

From an evolutionary standpoint, certain male characteristics should be critical to female choice (Halliday 1983). Although the importance of each variable changes with the phylogenetic, ecological, and social makeup of each species, it's possible to think through several attributes of males that we assume could be important.

The first thing a smart female animal should look for is a male of the same species and the right age. Some behaviorists think that the best thing a female can do is find a male with "good genes." Such a selection sounds like a reasonable expectation, and truth-in-advertising would certainly help females figure out who has good genes. But the argument for good genes is confusing because we define "good genes" as those that are passed on and, at the same time, suggest that good genes are the reason that an individual is reproductively successful in the first place. And no one seems to know exactly what "good genes" really are. Perhaps good genes are those that make for a strong immune system that resists diseases and parasites (Hamilton and Zuk 1982), or maybe good genes are those that help a future infant escape from predators. Unfortunately there's no clear way to define good genes, not even for our own species, because "good" is relative term that fluctuates with environmental conditions. Certainly no research on primate females has suggested that female primates have made specific choices based on a male's genetic quality. Females might also choose for male status or the goods he might provide, such as territory, food, or parenting.

It might also be advantageous for females to prefer familiar males over unknown males, and such a choice would be possible in groups in which males hold tenure for long periods or don't emigrate. But a penchant for the familiar might be overshadowed by a taste for the unknown—the novelty factor. Females are presumably interested in these males because they are different and attract attention. This selection for novel males may have evolved to avoid inbreeding. If a female is attracted to an unknown male, most likely he won't be related to her. A female might also find genetic variability for each conception when she seeks out novel males one year after the next.

And finally, we must consider the possibility of popularity. Humans base many of their mate decisions on "attractiveness." The components of beauty are distinctly different from culture to culture, and they change with time. We can't, however, dismiss the possibility that nonhuman primates, especially, have some concept of attractiveness, or beauty. In fact, many of the choices made by nonhuman primate females which seem unexplainable to human observers may fall into this category. One male in my group of Barbary macaques was repeatedly the target of female attention. To my human eyes he seemed a brute—he bit females and never engaged in mutual grooming. But the Barbary females saw in him "a certain something" only a macaque female could explain. His rate of copulation was higher than that of any other male.

If paternal ability, friendship, or unfamiliarity are important to females, recognizing individual males and their behavior would be paramount to the choosing females. This view suggests that unlike Darwin's females, described as dunderheads unable to deal with decisions, females are "socially intelligent" (Cheney, Seyfarth, and Smuts 1986, Small 1990a). They must have the ability to discriminate among males, repeatedly interact with the same males, and possess a large memory for sorting social information and understanding a social network. Because primates are known for just this type of mental processing, this suggestion seems reasonable.

What we don't know is whether female primates in particular actually *have* any of these proposed druthers. Researchers have painted a female choice world in which female primates have read the literature on what they "should" want. Our most difficult task is understanding what they *do* prefer, and exactly how they are motivated to make choices based on those preferences.

Female Choice among Primates

Primates, with their big brains, high intelligence, and flexible behavior, are likely candidates for studies of female choice. Whatever the theory, thirty years of primatology, including laboratory and long-term field studies on females, have shown that female primates are active participants in the mating game. They often initiate sexual bouts, and they walk away from male partners as well. Females are highly sexual, mating not just for conception at the moment of ovulation but repeatedly during the estrus cycle. They are clearly sexually assertive, but we still don't know what the evolutionary effect of that assertiveness might be, or why it evolved. And we aren't sure how a female primate's choice of a partner has shaped the evolution of primate mating systems.

To look for the original kind of female primate's choice, we have to step back and see if there's something to look for in the first place. Recall that Darwin and Fisher and others after him were trying to explain the appearance of odd traits in males. Let's broaden that to say we're trying to explain traits that appear in one sex and not the other. Did a trait come about because the opposite sex liked the trait, selected it over generations, and drove it to extreme proportions? Although this "Fisherian" type of female choice has been used to explain such obvious things as the peacock's tail, we're hard pressed to find such extreme sex-distinguishing traits in primates. Male primates are usually bigger than females, but this larger size is probably due to intense pressure on some males to compete with other males during mating. In other words, large male size has been sexually selected, but through male-male competition. Male baboons have hugh canines, but females have pretty big ones too, and the male exaggeration is also due to male-male fighting. There are only a few examples of male traits which might need some Fisherian female-choice explanation. For example, several lemur species come in sexually dichromatic colors (see Figure 3). The males are onle color and the females another. Because the color within a sex—male black lemurs are all black, whereas their female counterparts are all brown—is uniform, this trait probably didn't evolve to help males in fighting. It may have evolved by female choice, even if the color difference has nothing to do with male quality. Another possible example is the brightly painted red-and-blue-striped faces of male mandrills: just as with many bird species, only males have the bright faces while the females faces are dull brown. This coloration too is a possible avenue for female choice by Fisherian selection. One characteristic among primates has been clearly targeted for possible selection by Fisherian female choice—male penis size. Primate males living in groups with many females and many males, groups in which promiscuity is the mating rule, have long penes (Dixon 1978). Male chimps, in fact, use their penes for display toward estrous females. Because a longer penis would give a female pleasure (note that the human male has the longest and thickest penis of any primate), female choice might have been a factor driving penis length to extremes among primates. But examples such as these are few and far between, and no one has tested for the possibility of female choices in these specific examples.

In the second arena of female choice, primates should presumably excel—smart and discriminating females making smart choices for their own reproductive benefit. Sometimes these "Triverian" choices can affect the evolution of male traits, but no matter, female choices affect females. And these smart primates, with the largest brains in the animal kingdom, should be putting their selective powers to work.

How Primate Females Choose

If female primates have been selected to make mate choices, how can they exercise those choices? Females can't count sperm, do health checks, or predict male vigor and then point at the chosen one. They can only make decisions based on cues provided by males and then try to make some attempt to gain a particular male. The task for the behaviorist looking for indications of female preference is to discover exactly how females express that desire, if and when it appears. They can't tell us, and we must rely on what they do to indicate favoritism or refusal.

It's sometimes impossible to tease out preference initiated by the female. We never know what's going on in her head or even whether she's thinking about an array of males at all. How would we know if she prefers male A over male B unless she makes a move toward him? It's much easier, however, to spot refusals, which are also an indication of female preference and choice. She can run away, sit down when an unwanted male approaches, or otherwise refuse to cooperate—the monkey version of having a headache. Japanese macaque females frequently refuse males who approach them in a very solicitous manner; the male cowers up to the female, bobs his head in her direction, lip-smacks, and then makes his intentions clear by turning his rear toward her (Enomoto 1974), but he's often foiled because the female isn't interested in him, or perhaps the timing just isn't right. As Michael Huffman reports for Japanese macaque females, "Regardless of the male's persistent attempts to initiate a mounting series, the females had control of the interaction" (1991b, p. 107). Huffman's view comes from observing that 43 percent of all copulation solicitations by males were refused by females. African vervet monkey females are just as uncooperative. Sandra Andelman reports that males were successful only 42 percent of the time they tried to have a sexual interaction with a female. Vervet females "simply sat down or walked away from males," she writes (1987, p. 788). Sometimes, when a male persisted, female refusal escalated into all-out attack on the pestering male. The irritated female would bite, chase, or hit a male; she might also solicit female allies and mob the so-called suitor. These refusals are sometimes the best indicators of female preference that we have—a nonchoice.

The act of positive preference can start with a close association. After all, females are most likely to copulate with males in the near vicinity. Just walking up to a male and staying near him will increase a female's chances of mating with him, but although a behaviorist may see "close proximity" as indicating favoritism it doesn't really prove preference, and certainly not choice in the evolutionary sense. A better indicator of a female's preference is the sexual present. A female walks up to a male, presents her hindquarters, and effectively "chooses" him for a mounting (see Figure 4). Presents of this nature are rarely given by females outside of estrus. During non-breeding season, the hindquarters' present means "I am of lower status than you and here I acknowledge it," and this submissive gesture is given most often to other females. But during estrus, the same present is a sexual invitation, and it's our clearest indication of a female preference. What could be more direct than having a female rear pushed into the male's face? The data on several kinds of non-human primates show that females are instrumental in initiating copulatory sequences by approaching males and sexually presenting. Barbary macaque females, for example, initiate most of the copulations in this manner (Small 1990b). This is also true of rhesus macaques, Japanese macaques, langurs, and many other species (Blaffer Hrdy 1977, Huffman 1991b, Manson 1991, Wolfe 1979).

Presenting her rear in the face of a male is not the only way a female shows her interest. A female might walk up to a male, quietly sit at his side or close to his back and grimace. She might make the interaction more intimate by a simple touch, or she'll reach out to groom him. A more motivated macaque female may leap on her favorite male and rub her genital area back and forth on the male's lower back (Wolfe 1979). A lion-tailed macaque female will pester a male over and over, especially if he ignores her. She'll rush past him, yank his tail or hair, or even jump up and down and screech. In captivity, frustrated lion-tails have been seen writhing on the floor in front of a reluctant male (Lindburg, Shideler, and Fitch 1985). Female ruffed lemurs—black-and-white fluffy lemurs of the Malagasy forest—initiate copulations by approaching the male and slapping him (Foerg 1982). They might even beat him up to get his attention. On the other hand, if a male rudely tries to start a copulation without the female's first slap, he is beaten and chased away by the female. The same sort of female assertiveness occurs among capuchin monkeys in South America (Janson 1984). Females of this species chase the highest-ranking male, whine and whistle in his direction, and play a game of sexual tag. They run up, slap him lightly, and run away.

Even in species in which females seem overpowered by males, their subtle cues can influence what males get to do. Female Hamadryas baboons appear totally dominated by males (Kummer 1968), but females sometimes give nonconsort males subtle clues that let them know that the current partner is of no special importance. A nonconsort male will most often challenge an established consort and try to take possession of a female when he receives these sly clues from the female (Bachmann and Kummer 1980). Female Hanuman langurs are often terrorized by infanticidal males, yet the sequence of copulation is dictated by females. Langur females don't display any external signs of estrus—no swellings, no color changes. The only way a male langur knows a female is ready to mate is when she tells him so. The typical female langur in estrus walks up to a male, turns her back, crouches low, keeping her extra-long tail down, and bobs side to side in what primatologists call a "head shudder" (Blaffer Hrdy 1977, Sommer, Srivastava, and Borries in press). This dance of solicitation is repeated almost five times more often than responded to, and no one knows why males aren't always interested.

Females sometimes make their mate decisions under pressure. When a Japanese or rhesus macaque female sits next to a potential mate, she is often harassed by a higher-ranking male (Huffman 1987, Manson 1991). The unhappy male will run at the pair, and often attack the female; this is called a "consort intrusion." But the strategy seems useless for the attacking male. The pair almost always reunites, and interestingly enough, the female is always responsible for the reunion. Perhaps the low-ranking male is afraid to take the initiative after being attacked by one higher in status than himself. But even in the rare cases when the female stays away from her former chosen partner, the victimizer doesn't have it easy. He still has to court her in the usual manner by approaches and inspections.

Females may also vie for a certain male. Although sexual selection theory suggests that males rather than females should be expected to fight over mates, limited access to especially preferred males can also result in female-female competition (Robinson 1982, Small 1988). Several times I've seen a Barbary female displace another female sitting next to a male. In almost all cases, a higher-ranking female quietly walks up close to the couple and the lower-ranking female moves away from the male. The female who interrupted didn't necessarily mate with the male, but she did stop any further interaction between her lower-ranking troop mate and the male. This policy is probably not that effective in Barbarys, however, because the displaced female could just move along to the next male.

And finally, a female might be able to affect a mate choice by altering sperm transport internally after copulation (Cohen and McNaughton 1974, Overstreet and Katz 1977). This suggestion seems quite wild, but as William Eberhard points out when writing about females and conception, "Female genitalia may be designed not only to facilitate fertilization, but also to prevent it under certain circumstances," (1990, p. 138). Eberhard suggests that the human female cervix moves away from the vagina during intercourse, almost as if it were blocking the way for sperm to enter the tiny hole. Also a certain amount of sperm spills out after copulation; consider the spillage when a human female stands up after intercourse. Postural change that causes spillage isn't exactly passive acceptance of male deposits. Although monkeys and apes aren't bipedal, they can and do sit upright. Sitting up may be a subtle attempt to empty the male's sperm out of her body. Also, if the female is mating with multiple males during her period of sexuality, the intromission by one male pushes aside the ejaculate of another male. Males of some species have responded to multiple deposits by evolving a semen that coagulates and forms a plug. This plug functions to keep sperm from falling out, but it can't stop another male from moving in. Once females move on to another male, the first suitor is virtually helpless if he can't physically stop her. It's then her decision.

We can also document the importance of female preference some-times not by what females do but by what males do. Some primate males have courtship displays, although they are not all that common. Males perform this special pattern of behavior to attract females. The peacock's tail, and the way males use it, again, is the best example of a male display in a nonprimate. Primate males don't have such flamboyant options, but there's some evidence that males try. Their courtship displays are an indication of female power and confirmation that males don't run the mating game. For example, male chimps most often display to females in estrus to gain their attention (Goodall 1986, Tutin and McGinnis 1981). A male chimp doesn't just walk up and hop on a female. He patiently sits with his legs spread apart and waves his penis in her direction. He might also stare at her or shake branches at her. Because tension often runs high between male and female chimps, these displays tell a female that his interests are purely sexual in nature and that there's nothing to fear. Other primate males also resort to displays. When a mangeby male spots a female with a swelling, he tosses his heads rapidly back over one shoulder and the female often comes running (Wallis 1983). Japanese macaque males present their hindquarters to females. They sometimes attack females and threaten them as a sort of attention getter (Enomoto 1978, Huff-man 1991b, Takahata 1982a)—not much of a way to court a female but certainly an instant avenue to her consciousness. In all these cases, it's the female who has the final say. Males in these species don't force females, but must win them over. The decision is up to the fairer sex.

Theory Meets Real Life

With all this theorizing comes more confusion than clarity. Female primates, like all animals, are supposed to be choosy. And when they make mate choices, they should attend to the quality of the male and make smart choices. If this is true, the behavior of female primates should support the theory. If not—and my observations of the Barbary macaques do not support the theory—there's something wrong with the theory, or at least it doesn't have universal application.

The task at hand, then, is to determine who conceives with whom and what role female actions play in the conception of offspring with particular males. To what extent can the mate preferences of female primates be determined? How powerful are female primates in expressing their mating preferences? And do female preferences and choices have an evolutionary effect?

The next three chapters examine the reproductive physiology and sexual behavior of female primates, nonhuman and human, to show how theory meets real life, at least within the limitations of what we know about mating behavior today.

CHAPTER 5

Females Just Wanna Have Fun: Patterns of Sexual Behavior in Female Primates

WE HUMANS THINK about mating in two ways. For some (or on some occasions), mating is raw sex—two people attracted to each other engage in physical contact and sexual release. But more often mating is part of a relationship, a joining of two people with a connection, and perhaps even a future, in mind. The eroticism is therefore part of a larger whole that may include family and mutual finances. How is this mating similar to, or different from, the sexual behavior of the other primates, and what roles do females play in the staging?

Female primate sexual behavior doesn't begin and end with mate choice. Instead, a female's behavioral pattern is circumscribed by particular species-specific reproductive and sexual physiology. That physiology determines when she will reach reproductive age, what kind of cycles she'll experience, and whether she will have litters, twins, or singleton births. A female "should" be trying to improve her reproductive success within the bounds of her physiological makeup. For primates, we know much about the female cycle. We know the course of the cycle through changes in hormone levels, how conception occurs, and, in some cases, how many infants an average female might raise in a lifetime. Scientists are more uncertain about female primate sexuality. Although behaviorists have spent endless hours watching female nonhuman primates mate, there's still much to learn about who initiate copulations, which partners are preferred, and why certain matings occur and other options are bypassed.

This chapter focuses on female primate sexuality, not mating strategies, which I discuss in the next chapter. By "sexuality" I mean the sexual potential of a female, her sexual physiology, motivation, and pleasure. A female's sexuality is part of her larger pattern of reproductive behavior and defines which strategies she might be able to employ. For example, a female with a short period of fertility during a prescribed breeding season might be compelled to be promiscuous, while a female who cycles continuously until she conceives might be able to be more circumspect about mating. In other words, reproductive physiology and sexual potential are part of the total picture of female mating behavior and of the broader mating strategies that might or might not be open to each female.

Primate Reproductive Physiology

We begin with what primate females begin with—their inborn reproductive physiology. For each species, female physiology is constructed by millions of years of evolution. Not all the parts function perfectly, and there might have been a more efficient blueprint, but it's all we have. The particular reproductive physiology of a species is also its evolutionary baggage, the constraint within which each female, or male for that matter, must operate. Rather than explore why female primates have the reproductive physiology they do, this chapter takes the appearance as a given.

For decades, rhesus monkeys have been used in reproductive research. At the California Primate Research Center in Davis, California, thousands of female rhesus are housed in single cages, ladies-in-waiting for research projects on human reproduction. These monkeys have been trained to poke their butts toward laboratory technicians, who take a quick swab of the monkeys' vaginal areas each morning. Looking under the microscope, a technician examines the shape of cells taken from the vaginal walls and is able to tell how close a female is to ovulation. If she's close, a male monkey is introduced into her cage. This step in the breeding protocol can be tricky. What if the female takes an instant dislike to the male? The "chemistry" between a female and male macaque is just as mysterious as the chemistry between a woman and a man. Maybe she doesn't like his smell or the way he looks, and she backs into the corner and won't let him near her rear end. The male also has preferences. Like the female, he may be uninterested, or even repelled. If the technicians are lucky, the female likes the male, and vice versa. The monkeys may spend a few minutes grooming, and then the male grasps the female by her hindquarters, moves her into position close to the front of his body, and mounts her. He'll step up on the lower section of her back legs, grasp the fur of her rump and begin to thrust. An excited female often looks back at the male, and she might emit a loud bark. Males also often screech in excitement, a spontaneous call of pleasure. An ejaculating male pauses during the final moment, steps down, and may groom his partner for a few minutes.

The purpose of this single-cage pairing is to produce an infant, but there's more science to this process than a mere fertilization. With this method, researchers are able to time conception down to the day, and in studies of fetal growth, or of the effect of drugs on fetal development, the exact day of conception is a vital piece of information. Controlled tests during certain trimesters can then be administered relative to fetal growth stage. The point here is that these monkeys are used regularly in studies of fetal development because their reproductive system is so similar to human systems. People don't seem too scared when they hear that a particular drug harms rat embryos, but when the same chemical compound deforms monkey fetuses, the public is appropriately alarmed. These studies indicate something else to the evolutionarily minded: the similarity between human and nonhuman primate reproductive physiology, especially among the monkeys, apes and humans, tells us of a common ancestry. Although there are many differences between the reproductive process in monkeys and humans, we share a basic physiology.

The reproductive life of both human and nonhuman female primates is governed by hormones. Hormones are chemical compounds secreted directly into the blood stream (Leshner 1978). They act as signals for different cells in the body, telling them what to do, when, and at what rate. There's a simple feedback system from the tissue secreting the hormone and the tissue that's the target of the hormone—a direct telephone line between the signaler and the signalee with no interference and no party lines. When a hormone is secreted, it reaches the target tissue and causes a particular reaction and the target responds. This process goes on continually throughout the body, sometimes moving at a regular and predictable pace, sometimes dramatically affected by outside influences or other internal rhythms.

The hormonal process directing reproduction for female primates begins in the hypothalamus of the brain. The hypothalamus releases a chemical called gonadotropin-releasing factor (GnRH), which floats directly toward a small endocrine gland, the pituitary, located right between the eyes, suspended from the brain and set in the sphenoid bone of the skull (Moore 1988). This small gland, about the size of a pea, is a master switch in the hormonal circuitry of female reproduction. The anterior, or frontmost, lobe of the pituitary is dedicated to the female cycle. It's first job is to produce follicle-stimulating hormone, or FSH. Although following hormones through their maze of effects is often complex, the job is made easier by their names—most hormones are named after what they actually do. FSH, for example, stimulates follicles. When it's released into the female blood stream, it travels through most of the body, having no effect at all. But when FSH reaches the female ovary, something remarkable happens. The FSH molecules have found their target tissue, and the ovary responds by stimulating several follicles on it's surface. One of the follicles grows faster than the others, and the remainder stop growing. Within each of the follicles is an ovum, a female egg, and it begins to grow under stimulation of FSH. The eggs have been there ever since the first few months of the female's own fetal life, when she developed ovaries. She's carried them intact into her new life, a gift from her mother and father. She won't make any new eggs after she's born and she won't use up all the ones she came with. In addition, each ovary isn't exactly a packet of fully finished eggs. The eggs are cells left in limbo, arrested in the first phase of cell division, just waiting for FSH to come along and kick them out of torpor and into action.

As the follicle is encouraging the egg to awaken, the follicle wall itself thickens and a layer of connective tissue appears on the outer surface of the follicle. This outer layer then begins to produces another hormone, estrogen. This hormone gets the most press as the "female" hormone. To be sure, men also have estrogen (and women have testosterone, the "male" hormone), but the difference between the sexes is in the amount of each compound within the body and the exact time when it was released during fetal life. Females have more estrogen than males do, and this hormone is important to both reproduction and behavior. During female cycles, the estrogen produced by the stimulated follicles is dumped into the blood stream, and as each day progresses and the follicle grows larger, more estrogen is pumped into the blood stream. Estrogen flows upward through the body to the pituitary, day after day, raising the thermometer level that indicates how the follicle and its nestled egg are doing. Once the level of estrogen in the blood reaches a certain level, the pituitary "knows" the egg is about ready. The real process of ovulation is right on the threshold.

Ovulation, the release of an egg, also begins with a signal from the pituitary. Reacting to the level of circulating estrogen, the pituitary sends out a surge of another hormone, luteinizing hormone (LH). This chemical also has a direct effect on the ovaries, but this one acts as an explosive device. When LH surges from the pituitary, it races toward the ovary and literally bursts the swollen follicle and allows the egg to leave its protective custody. During ovulation, the egg completes its first cell division, staring the chromosomal journey toward reducing its chomosomal number to half. Only when the sperm enters the egg is the second phase of this division complete, and the egg ready to recombine with another set of chromosomes. Although ovaries aren't directly connected to the rest of the female reproductive tract, they are surrounded by long flagellating fingers that draw the egg toward the fallopian tubes. Like a spaceship with a tractor beam, the flagella wave the egg into the tubes. These tubes, about the same circumference as a strand of spaghetti, are the tunnels into the uterus. They're also the microenvironment within which fertilization occurs. But nothing happens instantaneously here. The egg takes several hours to reach the outer, and widest, section of the tube, the ampulla. If the female has mated, the entire tube may be filled with sperm, and it's here that conception will occur.

Popular mythology paints a picture of conception as an outdated allegory of human sexuality, with the male as aggressor, persuader, conqueror. The truth is far removed from that idea of hardy, brave sperm penetrating a passive unsuspecting egg (Freedman 1992). Instead, recent research on human fertilization has demonstrated that the sperm is actually an unwilling participant in the reproductive process. Sperm have spent their short lives trying *not* to stick to things. They move like grease-coated bullets through the male tubes and explode out of the penis at ejaculation. They race to the cervix, pass through a thick veil of cervical mucus, and try not to get caught in the crevices of the female uterus. Once they spill out into the fallopian tubes, sperm don't suddenly change their tactics and rush to drill into an egg. They are, instead, reluctant to contact anything, and it's the egg that must pull them toward fertilization. Researchers have discovered that a particular chemical excreted by the follicle of a mature egg actually sucks sperm in its direction (Garbers and Eisenbach 1991). The egg then reaches out and envelops reluctant sperm into its outer layer, the corona radiata. With their heads finally in direct contact with the egg, many sperm wiggle and release an enzyme from the tip of the sperm head. This enzyme helps break down the outer layer of the egg's coating, and one sperm is pulled in even deeper. At that time, the outer layer slams shut to all other sperm, and conception has occurred. This slight shift in perspective, from penetrating sperm to reluctant sperm, is a blatant example of how social norms and values dictate how scientists often perceive biology. If men must pursue women to implant their sperm (which is the popular notion), the egg must be a passive victim in the process as well. Contrary to popular belief, the egg is actually the aggressor. This isn't a sexual act, but it *is* reproduction, and we see that female biology even at the level of egg and sperm interaction, doesn't necessarily dictate a docile stance.

The Primal Primate Urge

Although ovulation and conception proceed in a similar manner among most primates, the sexual behavior of nonhuman and human females can be strikingly different. One of the presumed major differences between nonhuman and human primates is that humans don't have estrous cycles. The word "estrus" is used specifically in reference to nonhuman animals because sexual behavior among most female animals other than humans occurs only during a prescribed period that can be delineated from times when no sexual behavior occurs. Most female animals put sex into a behavioral box and take it out only at the appropriate time for mating, a time dictated by hormonal changes. For example, puppies and kittens appear in the spring and the fall, because female cats and dogs come into heat and mate only twice a year. For most primates, the period of heat, most often called estrus, is also usually, but not always, confined to a certain cyclical pattern that coincides with female fertility. When a female monkey is in estrus, she copulates. The rest of the time she's of no interest to males, and even if she were she'd be reluctant to mate. Most nonhuman primates' lives are so parceled out that sexual behavior has a specific time and place—and this system seems so efficient compared with the human case. For example, Hanuman langur females have a special mating dance they perform only when they want male sexual attention. A female scoots up to a male, sticks her hind end close to his body, and bobs her head up and down in invitation (Blaffer Hrdy 1977). Many female monkeys and apes, in fact, walk right up to males and swing their rears around as a signal that copulation should occur. Sometimes, these changes in behavior are the males' only cues that a female is ready to mate.

The length of estrus varies among species. Some females have extremely short periods of sexual interest, such as only one day for gorillas (Watts 1991), while other are interested for weeks, such as the ever-ready bonobos (Kano 1992). The stages of female estrus are so clear that Frank Beach, a comparative psychologist, has outlined three levels of behavior which, when observed together, are a good indication that a female is in estrus (1976). First, females must be "attractive," meaning that males are interested in them in a sexual way. A male baboon may often keep company with a nonestrous female; they're sort-of friends (Smuts 1985). A male grooms his favorite female, is tolerant of her infants, and he sits next to his female friend more often than with any other females. But this isn't a sexual relationship because the female isn't in estrus. When that same female comes into estrus, sprouting a large pink swelling, the male's behavior toward her changes dramatically. He'll try to form a consortship with her by sticking by her side all day. If she moves, he moves. If she eats, he eats. He also tries to keep her from other males by chasing away any troop mate that gets too close. Once her swelling has gone down, their friendship returns and the relationship becomes nonsexual once again.

Beach also suggests that we look for "receptivity" in females. The sexual meaning of "receptivity" is much more specific than the way we use the word in everyday language—to be receptive to a new idea.

Sexual receptivity in a female animal is the only time when she willingly allows males to copulate with her; a switch has been turned on inside the female's motivation circuit, and she's suddenly open to overtures from males. Females who are receptive stand in the right position for copulation and actually facilitate the mating. A confused male macaque mht walk up to a nonestrous female peacefully eating grass and push her to stand up on all fours. He wants to mate, but he can't do it if she lies on the grass. But she's not receptive, therefore she's not interested in standing up, and his attempt is thwarted. If the female had been in an estrous state, and thus receptive, she'd probably have stood up and the copulation bout might have proceeded.

The final criteria of Beach's definition of estrus is when a female is "proceptive," and this pattern of behavior is most useful for those studying female choice in mating systems. Proceptivity is female assertiveness toward males and mating. When females are proceptive they don't just passively wait around and allow males to copulate; they make moves all by themselves. The sexual present is the best indicator of proceptivity in primates. A female in estrus walks up to a male and presents her rear for inspection.

Beach suggests that estrus isn't a single one of these conditions—attractive, receptive, and proceptive—but that the combination of all three defines true estrus.

How human females fit into this scheme of estrus is unclear (see also Chapter 7). Human females don't restrict their sexual behavior to certain phases of the reproductive cycle. In fact, they copulate during pregnancy and lactation too. Do the conditions for estrus as outlined by Beach for other animals apply to humans? It's assumed that human females are almost always attractive to males. Not every day, not every woman, but in general there seems to be no predictable fluctuation of a man's attraction to a woman relative to a particular phase of her cycle. For humans, the condition of attraction probably has more to do with individual chemistry, cultural norms, and taste than with female hormones. Is receptivity in human females continuous or periodic? Although human females have been portrayed as continuously receptive (Lovejoy 1981), they really aren't, at least in Beach's sense of the definition. Human females aren't always ready to have sex (and males aren't either, popular literature aside). Potentially, human females might be able to copulate, but not to conceive, at any time—the vaginal canal is always open and a woman can be forced into copulation—but in Beach's strict sense, interest in sex for human females waxes and wanes. Some women report heightened interest midcycle or perimenstrually and less interest at other times (Adams, Gould, and Burt 1978, Udry, Morris, and Walker 1973), but so few studies have been conducted on human female sexual desire that there's no clear portrait of average sexual receptivity. As for proceptivity, women are so culturally bound to be passive sexual partners that it's almost impossible to know what the truth is. We can speculate, however, that women have the potential to be proceptive and receptive more often than the monkeys, who are more physiologically restricted by estrus, but in neither case are human females continuously in either condition.

Human and nonhuman female primates experience similar hormonal profiles over the cycle. The result of that hormonal profile is a monthly (or shorter or longer) cycle punctuated by ovulation midcycle. Human females cycle continuously if they are not pregnant or lactating, but many other primates have periods of down time during their reproductive histories. Some primates breed according to season. Many female macaques, for example, cycle only in the fall, and if they don't conceive they are "anestrous" or without a cycle at all until the next fall. Other species intersperse their cycles with pregnancy and lactation, regardless of the seasonal changes in their habitat (Lindburg 1987). The human cycle is called menstrual rather than estrus because sexual behavior isn't as tied to strict periodic events related to hormonal changes, and human females have a visible loss of blood and tissue at the end of each cycle. But the distinction between menstrual and estrous cycles isn't all that clear. Many nonhuman primate females have visible menstruation at the end of a cycle. I could often tell when the cycle of a Barbary female was over by the small trickle of blood draining from her vagina. Do Barbary females then have menstrual cycles intead of estrous cycles, or do they have both? More important, some nonhuman primates don't have estrous cycles with sharply delineated periods of sexual interest. Are their cycles then truely estrous if they aren't restricted to specific periods of sexuality? Bonobo females, for example, copulate all during their cycle and they seem always to be in estrus. Even when female primates do have more traditional estrous cycles, the period of female sexual interest is sometimes so long—for several weeks, for example—that it seems less hooked to hormonal changes than we might expect.

A Question of Swellings

Female nonhuman primates often use visual, behavioral, or olfactory cues to let males know they're ready to mate. For some species, the hormones of the cycle affect the special skin on the rear end or chest area of a female, creating a a video display of the changing chemicals within. A baboon female, for example, looks svelte from the rear until she begins to cycle. Slowly, the skin surrounding her anus and vaginal opening turns pink, then red, and it swells. This edematous tissue grows ever larger, day by day, like a bulbous tumor on her behind. Although the swelling appears burdensome to our human eyes, or at least an irritant, the female baboon goes about her business in the usual way. There's one major difference—the swelling acts like a lighthouse beacon that attracts males. Previous to the swelling, males weren't particularly interested in the comings and goings of this female, but suddenly, sporting the red flag of ovulation, she becomes the center of male attention—so much so, that males often come to blows over who will be her partner. Studies in the laboratory have shown that a few days after ovulation the swelling begins to deflate (Hendrickx and Kraemer 1969, Wildt et al. 1977). The swelling disappears as if pricked by a pin and the skin shrinks like a popped balloon. The female may be pregnant, or she may cycle again.

Some estrous females, including many strepsirhine primates, emit an odor in their urine, a pheromone, which males can't resist. Pheromones have less effect on primates than on other mammals because primates have a rather lousy sense of smell. But this connection between odor and female appeal has appeared in a series of studies of rhesus monkeys (Michael and Keverne 1970, Michael, Keverne, and Bonsall 1971). Nasal plugs were placed in the nostrils of a male placed in a cage with a female who had been swabbed on the rear with secretions taken from females at the height of estrus. At first the males didn't pay any attention to their partners. But when the plugs were removed and the males were once again placed to pair with the females in pseudoestrus, the males immediately began mounting, copulating, and ejaculating. What's interesting about this experiment is that the females were not actually in estrus, and thus the only clue the males reacted to was the borrowed secretions, the sexual pheromones, with their particularly odoriferous attractants. Male primates know all about these odoriferous changes because they spend time sniffing females or poking fingers into vaginal openings to gather secretions of willing females for a quick taste, their own personal monitoring device.

At first glance, it seems that outward displays of ovulatory readiness and direct solicitation are a major difference between nonhuman and human female primates. Certainly, human females don't have swellings, and we rarely thrust our hind ends into the faces of prospective mates. Are we then far removed reproductively from other animals? Not exactly. Detailed comparative research conducted to show how different we are from the other species has also shown some interesting similarities, or at least cast doubt on the depth of our disconnection from our primate cousins.

Sarah Blaffer Hrdy and Patricia Whitten combed through the research on primate female reproductive physiology and behavior (1987). They were interested in detailing the vast array of reproductive paths that female primates take—from those of the seasonal macaques to those of the harem-living gorillas. Although one can get lost in fascination with their seven-page table, a summation of the data is even more interesting. Much has been made about the large swellings on the rear ends of some female primates, such as the baboons I've just described. Anthropologists, particularly those trying to weave tales of human evolution, have used hundreds of pages of print explaining why human females "lost" those flamboyant displays (Blaffer Hrdy 1983). Blaffer Hrdy and Whitten's work has a surprising conclusion, however: the most common, perhaps the most basal and primitive, primate condition is without a swelling at all. Yes, baboons have swellings, and so do chimpanzees, but only 54 of the 78 species on which they could find information experience easily seen morphological changes during cycles. On closer examination, almost half (26) of those 54 showed only slight pinkness of vulvar or labial areas, a region pretty well hidden from all but the most intimate males. Large pink swellings, the kind that draw major attention, are few and far between, found in only 22 species (out of about 200 total primate species), and most of these species are the closely related macaques and baboons. In addition, our nearest relatives, the apes, are very inconsistent on this issue. Orangutans have no swellings at all; a slight whiteness appears on the labia during cycles of gibbons and gorillas, but only a true devotee could possibly see this change beneath all that hair. Only chimpanzee and bonobo females have extravagant swellings, and these two genera are closely related African apes.

Some scientists suggest that the important point is that our closest relatives, chimps and bonobos, have swellings. Inasmuch as we share common ancestor with the chimpanzee, that ancestor must have had a similar swelling and human females must have diverged from this path. Our prehuman female ancestors, the story goes, needed to stand up, and a huge swelling might have been in the way of a tucked-under genital area. Humans may also have been moving toward a monogamous mating system, and thus a female would only cause trouble by calling attention to herself and bringing a multitude of males together to argue over the right to her fertility (Alexander 1990, Alexander and Noonan 1979, Lovejoy 1981, Symons 1979). But a growing number of primatologists, influenced by Blaffer Hrdy's descriptions of the small distribution of primate swellings, have begun to take a different tack (1983). Perhaps the chimpanzee is the more derived species, they suggest. Under this alternate scheme, humans and the common ancestor would be more alike, and the female chimp may have been selected to evolve that swelling after humans and chimps separated. This idea is supported by the fact that the human and chimp lines separated first and then bonobos split off only 1.5 million years ago. Although bonobos have prominent swellings, the swellings never change much, so that bonobo females always give a signal of fertility—much as humans do. In addition, chimps and humans have such different mating and social systems. Female chimps mate with every male who asks, and a female with a swelling is the joy of a group of males. Her special status gains her attention and she might even travel with males. Although we'd like to use chimps as models for our distant forebears, we often forget that they evolved down their own path, shaped physically and behaviorally by pressures slightly different from those our ancient ancestors experienced. The chimp we see today is not our ancestor—it's our same-age cousin with a life of its own. In a sense, humans are more like bonobos, with a lack of any real cycle in sexual behavior. The point is, humans may not have lost anything at all (Blaffer Hrdy 1983). Our lack of a swelling may just be part of a shared ancestral history with all those other primate females who didn't have flamboyant swellings in the first place.

Whether or not humans lost anything, the appearance of swellings in other primate species requires explanation. Although no unifying principle seems to explain the appearance of these sexual signals in the smattering of primate species, there's an association between swellings and multimale social systems (Clutton-Brock and Harvey 1976). But the association isn't necessarily causal, for although swellings most often appear in females who traditionally live with many males, they are not typical of females in all species with multimale groups. William Hamilton suggests that swellings help males pinpoint ovulation, that their function is to aid males in not wasting their energy and sperm on less-than-fertile females (1984). But this explanation doesn't really work for females, and they're the ones with the swellings. In addition, swellings do not really pinpoint ovulation. They often last several weeks, and even the most trained eye can't tell when ovulation occurs until after the swelling begins to deflate, and for a male this would be too late. Swellings may be a generalized attractant for males, announcing her ripe reproductive state (Blaffer Hrdy and Whitten 1987, Lancaster 1979), but males don't attend just to swellings and color changes; they also need olfactory cues to respond to females sexually (Michael, Keverne, and Bonsall 1971). In addition, if swellings function to cue males, why did they evolve most often in multimale groups, where such attention would cause confusion and tension? Furthermore, if swellings are really necessary for females to attract males, why don't all species have them, especially solitary females who need to broadcast their fertility into the forest? It may be that swellings evolved to help females choose the best male. Swellings announce fertility and may promote male competition, and thus a female can sit back, watch the males fight over her, and choose the winner. Or maybe a swelling acts like a magnet that draws as many males as possible into a female's vicinity so that she has a smorgasbord of males to choose from. She could also opt to mate with as many males as she can attract to protect herself from male aggression and her infants from infanticide because males never kill their own infants. If all the males are possible fathers, no single male will risk killing an infant (Blaffer Hrdy and Whitten 1987). None of these proposals, however, provides a reasonable unifying explanation for the evolution of this large protuberance in response to hormonal changes, and we're left still questioning why some primate females have this flamboyant signal and some don't.

What Female Primates Really Do

When I studied the mating behavior of rhesus macaques and bonnet macaques in Davis, California, for two falls, I patiently watched females in an attempt to get a fix on how they went about mating. The rhesus were very disappointing—I rarely saw a copulation and didn't even get enough data to make comparative statistical tests among my ten subject females. The bonnets weren't much better, but at least they copulated a few times a day, often with a few different males. I assumed that together these two species painted the typical picture of macaque sexuality—a few copulations around midcycle, females sometimes moving among males, but more often vice versa. Possibly because they were caged animals they were shy, and most copulations occurred at night. After all, almost every adult female became pregnant without much sexual fervor. This experience didn't prepare me for what I would see several years later when I watched Barbary macaques.

When the fall breeding season began for the troop of Barbarys I watched in 1986, I could easily pick out the females in estrus: unlike rhesus and bonnets, Barbary's sport huge pink swellings signaling their fertility. I had read the literature about Barbarys—they had a reputation for being the most "promiscuous" primate, meaning that females copulated with as many males as they could, although exact data from a female's view were minimal (Taub 1980a). I saw two or three copulations an hour that fall, with females moving from one male to another (Small 1990b), and sometimes it seemed like an orgy, or sexual circus. The question that plagued me each day was, why this species and not others? What about their evolutionary history had pushed these females toward promiscuity? Back home, I began to search the literature for other cases of female "promiscuity" and discovered that Barbarys might not look so different from other primates after all.

My change in attitude about the nature of the sexual behavior of female primates echoes the historical understanding of primate females and their reproductive behavior. In many cases, female mating behavior was misconstrued because sexual behavior in these groups was studied mostly from the male perspective (Taub 1980a). Researchers believed, I think, that watching males would give a more accurate picture of the whole mating process. Or, researchers had biases about female primate sexuality. For example, in the 1940s Clarence Carpenter focused on rhesus females soon after they were released from holding cages onto the island (1942). The months of transportation from India and quarantine had had an odd effect on the females—as soon as they jumped onto their new island home, they came into estrus, as if release heralded the possibility of a new generation. More likely, they were finally free of the stress of closed captivity and their hormonal systems kicked into gear once they were out in the open again. Carpenter claimed that estrus for the rhesus was best defined as a period during which females sought out males; he outlined estrus as a complex of eleven behavioral and physiological changes. Carpenter's record of behaviors, although taken only during a two-month period, has become the blueprint for female rhesus sexual activity: all studies since then have been refinements of his early work (Conaway and Koford 1965, Kaufmann 1965, Loy 1971, Manson 1991, 1992). The conclusions Carpenter reached in a mere two months of observation are sometimes brilliant, but they have also proved wrong on several important points. Unfortunately, the flaws were often passed down though several decades of primate research along with the correct information. Carpenter is accurate in his basic description of the mating pattern. First the observer sees the physiological changes of estrus—a red face, a slightly swollen pink behind and legs. Females also actively seek out males, and when males copulate, they do so in a series of mounts ending in a final ejaculatory pause. During the series of mounts, males and females stay together, and they might form a consort pair and remain together for as many as a few days. Some of Carpenter's other observations, however, haven't stood the test of time. He thought that the increased activity by the female and her interest in males would encourage intense male fighting, and although males do bicker, the confrontations aren't that brutal. He also portrays rhesus females as out-of-control sexual maniacs—he says that females increase their general activity level when in estrus and get sleepy and sluggish once it is over, and that those wild and crazed sexual females are troublemakers. Estrous females, he says, are the "vortex" of stressful social relations. These needy rhesus females approach males and cause them to attack. Other females, he also claims, are offended by the new status of an estrous female and also attack her. Breeding season, by its very nature, is a more active time for both sexes, and when attacks do occur, males are most often responsible. Thus it's unfair to blame females for the level of aggression.

We now know that following females probably leads to more accurate data, and even more recently some of us have become convinced that one has to follow an estrous female continuously, all day, to see what she's really doing (J. Berard, personal communication). A female might spend the entire day "consorting" with one male, but the father is the one she met behind the tree when her "consort" partner wasn't looking. In addition, I suspect that human males and females haven't been particularly attuned to the possibility that females are copulating more often, and with more sexual assertiveness, than anyone suspected. Clutching the party line that females are choosy, behaviorists perhaps unconsciously overlooked the often less-than-choosy behavior of their subjects. No matter how deep the influence of human female sexual liberation on the studies of animal behavior, it seems that assertive sexuality, and the possibility of "promiscuity" in females, is still a difficult fact to accept.

Some of this revised picture of female sexual behavior comes from collecting long-term numbers on what females do, and some of the revision comes from recognizing that we are wearing human-oriented (or male-oriented) glasses when watching females mate. And some of the resistance to accepting what females really do comes from the dictates of evolutionary theory. Here is a primate female with one egg that needs to be fertilized. She has grown that egg from her small ovarian parcel and needs a father. Evolutionary theory predicts that this precious cargo should be treated like a crown jewel, to be shared only with the best male, the one who'll be a good genetic father or a good provider. Flying in the face of what they should do, some monkey, ape and human females flit from male to male as if they hadn't a reproductive care in the world. What's with these uppity females?

The least-promiscuous females should be those who mate for life. Such a female lives in a monogamous social system in which she is pair bonded to one male. This system presumably evolved because food resources are in small patches, and one male and one female are defending one patch. A male in this situation should be assured of paternity for all offspring born with his mate, and some primate males in monogamous systems care for infants more often than mothers do—an arrangement that makes evolutionary sense because the male is investing in a sure thing. But there's growing evidence that these males are often cuckolds: their supposedly bonded females are philandering. When a female leaves her pair-bonded male, as some South American titi monkeys do, and heads for a neighboring male, she might be trying to have the best of both worlds (Mason 1966). For some reason he's more attractive to her. Back at the home tree, the male waits patiently for his fertile and supposedly monogamous female while she mates with a neighboring male. When she returns, he may or may not be aware that she has copulated. Nonetheless, he'll care for the infant. Perhaps the chance of bringing up another's infant is so small that the resident male doesn't bother to stop the female or fight with the other male. Or he's willing to sacrifice her temporary unfaithfulness for a lifetime on a good territorial patch with a female. He may not father all the infants during the lifetime of his mate, but on average he'll do better than other males who play the field. Even siamang females, who barely copulate at all and are supposed to be strictly monogamous, apparently pick up males who aren't their mates (Palombit 1992). Again, males apparently do nothing to stop females and they sometimes go off themselves and find other temporary partners.

The same situation occurs in harem groups. For example, in almost half of the mountain gorilla harems, the female has access to another male. Along with the silverback, the chief of the group, a younger blackback male often tags along. The female mates with this male right under the nose of the silverback. But male tolerance in this situation may simply be a matter of familial love. Although male and female gorillas leave their natal groups, presumably pushed by primordial urges to avoid inbreeding, many males remain (Stewart and

Harcourt 1987). The silverback could then be the younger male's father or brother, and his permission to mate is not a matter of concealed jealously; the two males share genes in common, so why fight for a female? Or it may be that the silverback is tolerant because he's respecting the incest taboo and can't copulate with his daughter (Watts 1991). In any case, the gorilla harem isn't as closed as it appears.

Another "typical" harem among nonhuman primates isn't what it seems. For years, researchers had watched the fleet-footed patas monkeys on the savannahs of East Africa. These animals are considered the cheetahs of the primate world, with their long-legged racing across the plains. We used to think that patas lived in one-male harems and that males were very vigilant in guarding their females from predators. But there's a striking difference between the patas' social system and their mating system. Female patas come into heat for only two months each year. Their reproductive status, it appears, has a significant but temporary effect on the social system. After watching patas during breeding season, Dana Olson and Robert Harding discovered that outside males suddenly appear in the territory and compete for access to the group of fertile females (Harding and Olson 1986, Olson 1985). As many as eleven different males entered this particular study group and tried to copulate with females. Males fought with one another and interrupted copulations of others, and most males received visible injuries. On one day, the researchers write, there were ten receptive females and six males vying for them. As the females became pregnant or dropped out of estrus, the males began to disappear and the regular harem group reformed. For patas, the social system is a harem but the mating system is clearly something else: many males have access to many females, and females have a choice of many males.

Most often, female primates have several males to interact with and choose from, and they most often take advantage of the situation. In groups with multiple males, such as baboons, macaques, and many other primates, females most often copulate with more than one male, and sometimes several in rapid succession. This situation is most apparent in species with breeding seasons, so that several females are in estrus at the same time and no single male can sequester even one female all for himself. Promiscuity, or mating with every possible male, occurs most often when females cycle synchronously. Contrary to what theory suggests, these female primates are not selective and mate with more than one male even when ovulation is imminent. But I leave the details of this evidence for Chapter 6.

The Curious Case of Consortships

A major stumbling block to understanding primate female sexual behavior has been the limits we researchers place on their behavior. We label patterns of behavior and then have trouble when the label needs to be stretched and reformed. The notion of "consortship" is a classic case in primatology. In the early years, just by chance, researchers collected detailed mating data on savannah baboons. Baboon females, almost half the size of their male troop mates, are dominated by males, especially when large swellings indicate ripe estrus. Although a female may have the opportunity to mate with other males, an established consort partner keeps a steady eye on her and tries to make sure that she stays away from other males (see Figure 5). The word "consortship" was used initially to define the close male-female sexual bond seen in savannah baboons and then usage of the word spread to the relationship of other mating pairs. This semantic leap was a mistake. Researchers began to think that all primates form consortships, and they applied the word to any short or long, exclusive or nonexclusive mating (Small 1990c). What was originally intended to describe a specific male-female association that lasted during the days surrounding ovulation became an all-inclusive word for mating. In fact, in most so-called consortships females and males mate with others even if they happen to keep company for a few hours during the day or for several days, and very few mated pairs in multimale-multifemale groups stay together longer than an hour or so (Small 1990c). Again, the jargon gets in the way of understanding exactly how the mating process occurs. And this label has been especially damaging to the study of females—once a female is described as "being in consort," no one sees the importance of her regular copulations with other males. The word "consort" implies male dominance of the mating situation, and it doesn't seem to matter that females often wander away and mate with other males or that a consortship shorter than the window of ovulatory opportunity isn't much of an exclusive arrangement. For example, when a Japanese macaque female consorts for three days, she is highly likely to find an additional male and sneak off and mate with him (Huffman 1991b). Even this fact isn't especially relevant, because almost all consortships in this species last only one day. A longtailed macaque female will regularly mate with other males during her association with one primary partner (van Noordwijk 1985). Female blue monkeys in the forests of Africa initiate most of their consortships, but these partnerships may be in name only for the females involved. Researchers have discovered, by simultaneously watching all the parties concerned, that when a current mate is away chasing a potential male intruder, the female skips off and copulates with yet a third male (Tsingalia and Rowell 1984). Even those models of consortships, the savannah baboons, are not as perfect as they appear. Females consort with more than one male every time they come into estrus (Berkovitch 1991, Hausfater 1975), and female baboons are responsible for initiating 75 percent of these couplings. It may be that researchers find it difficult to accept the fact that males aren't necessarily running the mating game and that, even when there's some bonding between the sexes, females are often as likely as males to slip out and copulate with others.

Consortships are therefore something of a myth. The word implies a long-term exclusive mating, but we're hard pressed to find such a relationship in nature among our primate cousins. This lack of clarity on the consort issue is especially important to us humans because it has something to do with how we see our heritage. If no nonhuman primates are very exclusive in their mating practices, what does this say about human behavior? Is monogamy, something our culture treasures, only socially imposed?

In sum, one of the remarkable facts of primate mating behavior is its variability. Some species, including gibbons, the small apes, have rarely been seen copulating in the wild. Although researchers have spent endless hours craning their necks toward the highest treetops in southeast Asia, gibbons just won't be caught "doing it." Other primates are also on the low end of the libido scale—gorillas, for example, are the quickest, with a copulation lasting only eighty seconds (Watts 1991). At the other end of the continuum are Barbarys and two of our closest living relatives, chimpanzees and bonobos.

Females in these species sometimes mate with every possible male in the group, and they copulate much more often than is necessary for conception. Consortships do occur, but females and males also wander in and out of this aggreement and mate with others. The less-than-exclusive behavior of females is opposite to the behavior that evolutionary theory predicts, and so it begs for radical thinking about the evolution of female sexual behavior.

Orgasm, la Petite Mort

Swelling or no swelling, estrous cycle or menstrual cycle, female primates engage in sexual activities, and they seem to be motivated to do so. Females "ought" to engage in sexual intercourse only to conceive, but if this were true, females would copulate a lot less often and with fewer partners. What then is the motivational secret behind female primate sexual interest?

We do know that men have orgasms. They have to—without orgasms, there'd be no ejaculation of seminal fluid and males would not pass on genes to the next generation. Natural selection has favored a catapult orgasm system for males of most species, with the penis as the structure of trajectory and sperm as the ammunition. Evolutionary biologists assume that the penis, and forceful ejaculation, evolved because males who place sperm high up in female reproductive tracts probably father the most offspring (Parker 1984). But what about females? Many years ago, there wasn't any scientific proof that human females had orgasms, save for the descriptions of women themselves. But work by Masters and Johnson (1965a, 1965b, 1966) helped scientists confirm what women already know—female orgasm does exist, thank you very much.

The clitoris is the main organ of arousal and pleasure for human females (Gebhard and Johnson 1979, Hite 1976, Kinsey et al. 1953, Sherfey 1966). Using wires that monitor physiological responses such as changes in muscle tension and heart rate, sex researchers have found that there are four arbitrary stages of human female sexual response—arbitrary because the stages can't always be delineated and women may not experience each stage or may not experience the stages in the same order. The stages are excitement, plateau, orgasm, and resolution. The excitement phase is initiated by visual, physical, or psychic stimuli: women can become interested in having sex because they see something that piques their interest, because they are physically touched, or maybe because they are just thinking about something sexual. The plateau stage can be measured physiologically—the breasts swell a bit and the genital area swells as well. Masters and Johnson learned that women can rapidly and easily move from this second stage to the third stage—orgasm, and once at the orgasmic peak, women can fall back and forth between plateau and orgasm many times. Unlike males, who experience only one huge uncontrollable rush to orgasm and ejaculation, women can experience multiple orgasms. The female orgasm involves involuntary contractions of the entire genital area including the outer third of the vagina, the rectum, and the lower abdomen (1965a, 1965b). Although the clitoris is the initiator of the stimulation, the orgasm is manifest in the vagina and the rest of the pelvic area. Human females vary in the intensity of their orgasms; they range from casual to forceful. Women also don't have orgasms as often as men relative to the number of sexual acts, but the fact that female subjects of sexual research were easily able to bring themselves to orgasm suggests that relying on a partner may lower a woman's chance of reaching orgasm.

Given that women do have orgasms, *why* do they? Some researchers believe that the clitoris and female orgasm are atavistic traits, features left over, unnecessary, something passed over by evolution which serves no purpose. Donald Symons, an anthropologist, suggests that the clitoris in particular is a morphological gift resulting from a shared embryonic connection with the male penis (Gould 1987, Symons 1979). There's reason to support this idea. The clitoris is made of the same tissue as the penis, and it responds sexually in a similar manner (Masters and Johnson 1965b). Also, female sexual arousal, orgasm, or any pleasure isn't *necessary* for to conception the way ejaculation and orgasm are necessary for males to pass on genes. Human female sexuality, according to this view, then evolved something like this: the male evolved a penis, ejaculation, and orgasms for sperm transport, and the female acquired a clitoris and the capacity for orgasm as so much common baggage.

And yet, as women read this, they know something's askew; men who enjoy sex with women also know something's wrong with this approach. Surely, something as prevalent as female orgasm must have some adaptive value to females. Yes, females can conceive without orgasm, but they rarely copulate without being somehow aroused. And wouldn't evolution favor females who willingly engaged in sex to conceive? If females were sexually dead, chances are there'd be fewer babies on this planet. And there's something else—a sneaking suspicion that male scientists are trying to take credit for female sexuality, as if they, not evolution, were responsible for handing us the clitoris on a platter.

The males-got-it-first orgasm theory is shaky on several levels. First and most fundamental, the clitoris didn't suddenly appear as some atavistic trait only in the human female—it's found in reptiles, birds, and is most consistent in mammals (MacFarland 1976). It evolved when the urinary opening separated into distinct parts from the more primitive cloaca, the all-purpose orifice, such as the reproductive system now found in most birds. At this time, the clitoris formed in females as did the penis in males, and both have a small bone (the *os bacculum* in males and the *os clitoridis* in females) that signals their joint evolutionary history. Most primate females, including human females, are missing this bone, and human males have derived a penis that is also boneless. Nor is the clitoris simply a homolog of the penis. The blueprint of every fetus is essentially either female or sexually undifferentiated. It's not until week seven that under the influence of a Y chromosome the fetus takes a male direction (Moore 1988). In addition, the penis isn't really necessary for sperm transport. The male appendage isn't found in all animals (most birds, for example, don't have penes). In addition, some suggest that the female orgasm is not exactly like the male orgasm, and thus it would require its own evolutionary explanation (Konner 1990). The better question might be, why is sexual pleasure associated with reproduction for both males and females? Why do we humans want, need, and love sex? And is this pleasure part of our ancestral, primate female nature?

Ask most female primatologists if monkeys have sexual pleasure or orgasms and they'll respond, "Of course!" Those who watch the sexual behavior of our closest relatives know that the roots of sexual pleasure are deep. I've observed thousands of copulations in macaques, and I would say, yes, there's no doubt that female monkeys often reach orgasm. I couldn't tell you if those monkeys experience their orgasms in the same way a human female does, but then I also couldn't tell you if the orgasms of my female friends are like my own. I can only observe in my monkey subjects the body postures and vocalizations that suggest to me they are in some sort of passionate trance. Without a doubt, I'd say they are having a good time.

In 1971, Frances Burton published the first paper that substantiated the monkey orgasm. In a rather unusual research protocol, Burton rigged up three female rhesus monkeys in dog harnesses and used cat collars on their feet and hands to keep them strapped, but not immobile, on a table (Burton 1971). The heartbeat of each female was monitored by wires attached to her back. Burton then fed them, groomed each female for a few minutes to calm her down and bring monkey heartbeats back to normal after the trauma of the moment. She then stimulated the female's clitoris for five minutes. The subjects sometimes protested and tried to get out of the harness, but more often they were in estrus and thus cooperative once the grooming and stimulation began. A fake macaque penis was inserted into the vagina and thrust, as in copulation, for five minutes. Burton was trying to prove that rhesus macaques had orgasms, and she did find evidence for at least three of the four female response stages suggested by Masters and Johnson's work on human females. The monkeys became excited—during clitoral stimulation the vaginal opening dilated, vaginal mucus was produced, the labia enlarged, and the whole genital area deepened in color. Female monkeys also hit a plateau stage. The clitoris became swollen, the vaginal barrel widened, and the monkeys reached back to clutch the pseudo-partner. Burton believed that this clutching response was analogous to orgasm, as Doris Zumpe and Richard Michael had proposed a few years earlier (1968). Although she couldn't prove it, monkey orgasms seem to be signaled by the fourth stage of spasmodic body clutching and turning toward the partner. Her subjects also experienced a resolution phase. The genital area decreased its swelling and the coloration subsided.

Burton's work was followed up by Suzanne Chevalier-Skolnikoff and her observations on stumptail macaques (1974, 1975). The choice of this species was important. Rhesus monkeys mount in a series; that is, males mount for a few seconds, pull out, groom a female or just sit next to her, mount again, over and over until he finally penetrates for a longer time and reaches an ejaculatory pause. Stumptail males mount only once, but the copulation lasts for a long time. If females are going to reach orgasm, chances are they'd gain heightened levels of sexual excitement during these longer uninterrupted bouts. After watching five females and two males for over a year, and writing down the details of 143 sexual encounters, Chevalier-Skolnikoff wrote that orgasms of females were much like the orgasms of males. During copulation, females looked back at males and gave a "positive emotional expression." They also lip-smacked—a definitive happy face for stumptail macaques—puckered their lips into an "O," made breathy pants, and experienced body spasms. These are the same orgasmic behaviors that male stumptails exhibit at ejaculation. On twenth-three occasions, the female stumptails engaged in homosexual interactions, and they acted much like males mounting a female partner. Mounting females made the same number of thrusts as males by rubbing their clitorises high on the backs of partners, and they mimicked the ejaculatory pause. Chevalier-Skolnikoff points out that the sexual peak for these female monkeys was also much like that of humans. They have muscular spasms, their faces twist into a grimace, and their vaginas contract spasmodically. These same female behaviors have been noted with five species of macaques (Chevalier-Skolnikoff 1975). Chevalier-Skolnikoff's observations were reinforced by David Goldfoot and colleagues for six female stumptails (1980). In this study, the researchers inserted transmitters in the females' vaginas to measure uterine contractions. By watching needles jump on an EEG machine during homosexual mounts, they confirmed with real numbers that stumptails have vaginal contractions during copulation.

Although these invasive lab studies demonstrate that female non-human primates have orgasms and sexual pleasure, it's disturbing to realize how far researchers have gone to support their contention. Underlying this search for female sexual Truth is the ugly suspicion that when human females talked about their own sexuality, people had not believed them. We accept the fact that men have orgasms and that they have an urge for sex, but female sexuality in humans and animals, had to be proved.

Other, noninvasive, evidence speaks for the high degree of pleasure female primates experience in sexual activity, whether or not they have humanlike orgasms. South American spider monkeys and bonobo females have huge clitorises. Bonobos rub the constantly swollen organ on other females and the backs of males, and their behavior has nothing to do with reproduction. Some Japanese macaque females are more highly sexually motivated than others, and their motivation shows up in the way they go after males. In 30 percent of the copulations recorded by Linda M. Wolfe, females mounted a male and rubbed her genital area on his back. The females who regularly initiated sex with males by hopping on their backs also had significantly more male partners during the breeding season (1979). In a study of sexual behavior among sooty mangabeys, a study focusing on males, Deborah Gust and Tom Gordon discovered that 10 percent of the females manually stimulated themselves during copulations (1991). A female might reach back between her legs and repeatedly touch the lower portion of her genitals while the male is thrusting. Although only three out of the thirty-one mangabey females did so, those three did so regularly. Most captive female monkeys and apes have been seen masturbating, but observers often believe that this behavior is a pathological result of captivity rather than a normal activity. However, masturbation has also been seen under field conditions.

Even the uninitiated can identify female nonhuman primate orgasm, or sexual pleasure. Body postures of females differ from those of normal day-to-day social behavior, such as the clutch reflex in the macaques. Females also make noises not heard in any other context but mating (Hamilton and Arrowood 1978). During copulations, female Barbary macaques both clutch and call. The female turns her head toward the male and reaches back with an arm, grabbing at the flank of the male. But this standard macaque gesture is not as surprising as the distinct female mating call. Female Barbary macaques almost always emit a loud raucous call during copulation. It begins deep in the body cavity, a pant-pant sucked in and out at a rapid pace, rising to a chorus the entire group can hear. I remember standing more than fifty yards into a small forest and hearing a female call. I, and all the monkeys in my proximity, turned in the direction of the noise and easily identified it as the mating call of female 38. As we all went back to the business at hand, me to my notebook and the monkeys to eating, I realized how individualized each female's call was, and if I could identify a call individually after only one breeding season, surely each group member knows who's mating and when.

Ute van den Bergh and Jürgen Lehmann, who both believe that their Barbary macaque females have orgasms, studied the function of the female mating call during the 1986 breeding season. They lugged around giant tape recorders and waved wand-shaped directional microphones in the direction of calling females. It may be that such loud calls cause some sort of male-male competition, but there is no evidence that jealous males rush in to stop a screeching female and her partner. In the analysis of their four hundred recorded calls, they found an association between the call and male ejaculation (1990). Males most often ejaculate when females call. When there was a copulation but no ejaculation, the female had probably not made a peep. Van den Bergh and Lehmann conclude that the female calls encourage males to ejaculate, as if females are urging them on. The same male-female calling interaction has been seen in baboons (Saayman 1970). It's clear that the calls indicate high excitement for the females. For Barbarys, then, the females and males are working together like a well-oiled reproductive machine; her excitement spurs him on and the copulation isn't wasted. She can then move on to another male. Copulation calls by females have been heard in at least three species of macaques (Lindburg 1990), one baboon species (Saayman 1970), and chimps (Hasegawa and Hiraiwa-Hasegawa 1990, Hauser 1990). These sounds range from an o-o-o to a barking squeak, and while the authors have suggested that making a noise while having sex in the midst of a group may have other social functions, they also agree that the sounds suggest some sort of sexual excitement.

Homosexual Pleasures: Choosing Females over Males

Perhaps the best indicator of female sexual urges is not how females behave with males but how they behave with other females. Some might suggest that females copulate with males primarily to conceive and that thus homosexual behavior is aberrant, pathological, and crazy. Homosexuality requires no explanation, they'd say, because it's not the norm. In fact, female homosexual behavior is found in a number of species of nonhuman primates (Akers and Conaway 1979, Chevalier-Skolnikoff 1976, Sommer 1988). Sometimes female homosexual behavior is part of a long-term friendship or a sign of growing attachment between females forming an alliance. It can also be purely sexual in nature.

Many primate groups are female-bonded (Wrangham 1980), meaning that females have tight social relationships. They never leave their home groups, and there is social power in the alliances among females who've known each other all their lives. Most of the female bonds in these groups are formed along the lines of kinship. Mothers and daughters, sisters and aunts, hang together in a matriline and form a stable core. But of interest to the understanding of homosexual behavior among females are those species in which females form bonds that are not kin based. And at least among bonobos, sexuality is an integral part of making those nonkin-based female-female relationships work.

A genital-genital rub, better known as the "G-G" rub by observers, is the most frequent behavior used by bonobo females to reinforce relationships or relieve tension among females (Kano 1980, Thompson-Handler, Malenky, and Badrian 1984, White and Thompson-Handler in press). One female bonobo rolls on her back and extends her arms and legs. The other female mounts her and they smack their swellings right and left for fifteen seconds, massaging their swollen clitorises against each other. G-G rubs occur around food because food causes tension and excitement, but the intimate contact has the effect of making close friends. And sometimes females would rather G-G rub with each other than copulate with a male. In a short film about the bonobo group at the San Diego Zoo, bonobo females Loretta and Lina graphically demonstrate that males are sometimes the least-favored sexual partner (Small 1992c). In this ape short subject, male Vernon repeatedly solicits the two females, Loretta and Lina, by arching his back and displaying his erect penis. The females continue to move away from him, tactfully turning him down until they creep behind a tree and G-G rub with each other for a few seconds. They have a long-standing friendship to maintain, and perhaps to them a G-G rub feels much better than copulation with a male.

Female bonobos, who emigrate from their homes and are thus not bonded to other females in the group, also use sexual activity with other females to aid in their tricky transfers into a new community (White and Thompson-Handler in press). Bonobos are almost always sexually receptive, and transferring females systematically have sex-

Figure 1. Most primate infants are born into a social whirl (Chapter 2). Here two male Barbary macaques greet a new infant. (All photographs by the author, unless otherwise credited.)

Figure 2. The females of some species of primates, such as the longtail macaque, remain in their natal group while males emigrate. These females form tight alliances with their matrilineal kin, and they favor their female relatives over nonrelatives (Chapter 3).

Figure 3. Female primates may choose for specific male features, with the result that male characteristics become exaggerated over time. For example, the difference in coloration between these black lemurs (females are brown and males are black) may have evolved by female choice for darker males (Chapter 4).

Figure 4. Females have various ways of expressing choice. Here a female Barbary macaque presents her sexual swelling to a male. Not all female primates have swellings that advertise sexual receptivity (Chapter 4).

Figure 5. Males and females of a few primate species form consortships during the female's receptive period. An olive baboon male tries to maintain exclusive access to a female in estrus, but females are also instrumental in determining consort overthrows (Chapter 5).

Figure 6. Because primate infants are relatively dependent, a female's mate choice might be based on fathering ability. So far, however, there is no clear evidence that females make choices based on a male's past paternal performance or on his future potential. This male Barbary macaque cares for an infant, but his attention does not gain him female favors (Chapter 6).

Figure 7. Rhesus macaque females have a penchant for variety. They mate with high-ranking familiar males but also seek out low-ranking males peripheral to the group (Chapter 6; photograph courtesy of Elizabeth Wardell).

Figure 8. Bonobos are perhaps the most sexual of primates. Females use sex to gain food, transfer into new groups, and form alliances. They engage in sexual activity with males and with females (Chapter 6; photograph courtesy of Frans B. M. de Waal).

ual contact with each member of a new group, especially the resident females. Different from human or chimpanzee groups, in which unrelated females construct friendships by the ordinary roots of shopping or grooming together, bonobos do so sexually. Although pleasure may be the motivator behind this type of female-female assignation, the function is to form an alliance. A female bonobo solicits and gives G-G rubs at any opportunity, trying like an eager newcomer to make friends. These female alliances are serious business because they also determine the pecking order at food sites. Females with powerful friends eat first, and subordinate females may not get any food at all if the resource is small; when times are rough, it pays to have close female friends.

Another kind of female homosexuality seen in nonhuman primates appears to be related mostly to pleasure or to frustration with a male-monkey shortage. Although many researchers early on noted instances of female homosexuality among their animals (Carpenter 1942, Chevalier-Skolnikoff 1974), no one really tried to understand why females were having sexual interactions with other females until the late 1970s. In 1979 Linda M. Wolfe, a primatologist, brought female homosexuality out of the closet when she described the behavior of Japanese macaques (Hanby, Robertson, and Phoenix 1971, Wolfe 1979, Wolfe 1984, Wolfe 1986). She found that female-female mounts follow a particular pattern: they occur only during breeding season, when females are in estrus and sexually receptive; females use a mounting posture identical to that of males; females form typical consorts with other females as they sit together, groom each other, and forage together; there's an incest taboo among females—they don't form female-female consorts with close relatives. More surprising is the rate of homosexual interactions among her macaques. During one season, 45 out of 58 females engaged in homosexual activity, but this number decreased the next year. Wolfe even describes some females as "bisexual" or exclusively "homosexual." There seems to be a relationship between age and homosexuality—young females have the homosexual interactions most. Wolfe explains that making a sexual liaison with another female rather than with a male is perhaps easier for these youngsters. But this reasoning doesn't explain why so many prime-age Japanese macaque females habitually sexually engage with other females. The animals Wolfe studied had been transported from their native Japan to a site in Texas, and Wolfe decided that female homosexual behavior was a result of a skewed sex ratio that occurred when the group was moved to their new home: There weren't enough males to go around and females were turning to other females for satisfaction. But her later research, a comparison of monkeys in Japan with the Texan group, also showed that females in their native land regularly have sexual interactions with other females. Harold Gouzoules and Robert Goy (Fedigan and Gouzoules 1978, Gouzoules and Goy 1983), working with captive Japanese macaques, substantiated Wolfe's work and provided a physiological explanation for the mounting frenzy. They maintain that mounts occur as a result of a hormonal storm in these females, and the storm rages most when females are ovulating or pregnant. They found two peaks in mounting, both homosexual and heterosexual: when females are most fertile and sixty days after conception, both periods that correspond to dramatic hormonal changes. In other words, there's little social function in these liaisons; mounting is mostly the result of hormonal motivation.

Although some suggest that homosexuality isn't as frequent with other nonhuman primate species, it does occur. Arun Srivastava and colleagues utilized the data from six years of watching Indian langur females to theorize about langur female homosexuality (1991). During more than 3,000 hours of observation, they spotted 524 female-female mounts, and all the females in the study were involved at least once. Again, the posture and movement of the partners were strikingly similar to heterosexual mounting and copulation, although they claim that the clitoris of the mounter is not stimulated during thrusting and there was no sign of orgasm for either partner. The langur observers believe that female motivation for homosexual mounting is pure sexual arousal based on hormonal triggers. But these authors also point out that previous explanations for female homosexuality have addressed only the functional or proximate level of explanation—females mount each other because they are sexually aroused, want to make social ties, or are expressing dominance. They offer a more evolutionary explanation for the pattern of female homosexuality. Perhaps females are mounting others, they suggest, to decrease sexual activity by rivals: if a female copulates with another female, the partner won't go after a male, she'll be satisfied, with what they call "surrogate sexual satisfaction." This is a good explanation for the mounter, but it doesn't explain the cooperation of the mountee, a female who's in estrus and needs sperm to conceive. Perhaps the mountee is simply intensely motivated for sex with any partner.

Females sometimes get their pleasure from other females, and like all physical interactions among social primates, these contacts are often, but not always, meaningful in a social context. From an evolutionary standpoint, a homosexual interaction makes sense only when it is linked to a positive social context, such as making an alliance with another female. Although it's possible that females initiate these interactions to stop rival females from copulating with males, this effect seems secondary. After all, homosexual mounts can occur during all phases of the cycle and, at least among bonobos, during infertile periods. Harder to accept, perhaps, is the picture of nonhuman primate females that are so indiscriminate that they will mount anything that moves, or allow just about anything to mount them. Homosexuality among female nonhuman primates, may, in fact, be just part of a larger portrait of a highly sexual, even promiscuous, and not particularly choosy animal.

The Purpose of Female Pleasure

Taken together, these studies tell us that female nonhuman primates experience sexual pleasure, that some have the capacity for orgasms, and that they are motivated to engage in sexual interactions with males—or females. But the fact that human and nonhuman primates are motivated by sexual pleasure raises, of course, the obvious evolutionary question: what is the selective advantage to females? If pleasure and orgasms aren't necessary for conception, why should natural or sexual selection favor this trait as a motivator for sex?

Sarah Blaffer Hrdy points out that natural selection would hardly treat female sexuality any less seriously than male sexuality (1981a). Females are engaged in the sexual process as intimately as are males, and to be a cooperative partner surely a female must have a sexual interest as strong as a male's. What better motivator for sex than orgasm and pleasure? Females who mate often and under their own steam of needing sexual release can move among males and not be passive agents in their own reproductive futures. Other social benefits must also accrue while choosing and mating with males. Females might bond with males who then provide protection against predators or angry troop mates. Sexual females might, in rare instances, also gain food. Although some might call this selling goods for services, prostitution of the forest, there's no reason that a female primate shouldn't use her sexuality to further her future.

Some kind of sexual pleasure is something we human females share with other female primates and it's one reason female primates seek out sex. But human females seem to have taken orgasmic pleasure further than their primate sisters have (see also Chapter 7). Our orgasms are easily measured, women are able to describe the heights of the orgasmic experience, and women regularly seek out orgasm as part of sexual interaction. For humans, in other words, orgasm is a major feature of female sexuality, a driving force leading us to have sex and reproduce. For other female primates orgasm is sometimes hard to document and copulation often occurs so quickly that even if these primates have the capacity to reach orgasm there just wouldn't be time. For us humans, who lack the undeniable hormonal drive of estrus, which would push us to copulate regardless of our own pleasure, orgasm may have been selected in our species specifically to lead us into sex and reproduction. I suggest that orgasm for human females is the necessary feature of our reproductive biology which evolved to take the place of estrus. Without the strong potential for orgasm, human females would never be motivated to copulate at all. The fact that many women do not always reach orgasm reinforces the notion that it's a stimulus for females to engage in sex (Blaffer Hrdy 1979). Like rats in a maze who have been trained to push a bar to receive a food pellet that appears every once in a while, female primates seek out sexual interactions that produce the peak pleasure, even if it happens irregularly.

Thus selection for female sexual assertiveness, sexual pleasure, orgasms, and the clitoris as an organ of pleasure occurred parallel to, not as a by-product of, selection for male sexuality (Blaffer Hrdy 1979, Blaffer Hrdy 1981a, Fedigan 1986). Anticipated sexual pleasure for *both* sexes is most likely an initiator (a stimulus of sorts) to begin mating behavior.

The New Primate Female

We end up with a naturally sexual female primate. She's motivated to have sex and enjoys it with her special organ designed solely for this purpose, the clitoris. And it gets even better, depending on your perspective. Female nonhuman primates not only like sex, and are driven to it, but as Chapter 6 will show, they often mate with many different males and certainly more often than is directly necessary for conception. This brings us back to the issue of female choice. How can there be choice and yet promiscuity? Why would selection opt for a female primate who enjoys sex and seeks out orgasm but also require her to be choosy?

The Simian Times

PERSONALS

SM, dark gray rhesus, loves multiple mounts, fun guy, sperm now available.

Females with Swellings, want to be friends? Baboon male hopes to join your group and form consort.

MM, buff-colored gibbon, looking for extra-pair matings, good at defending patches, resonating long call.

Solitary Male, galago who must spend time alone hopes for sometime companion with similar needs.

Leapin' Lemurs, find me a gal who likes to leap like I do and we'll make a ringtail twosome. Ready to transfer to your group.

SM, Barbary macaque, will carry around kids for several months, likes assertive females, not the jealous type.

Swingers of All Sexes and Ages—join our bonobo sharing group, meet new folks, free-for-all fun.

SM, Latino titi monkey, enjoys caring for offspring, looking for lifetime commitment.

Single with Experience, blackback male gorilla, 4 yrs with previous harem, looking for a few good females.

Sub-adult Male, orangutan, won't force you, but would like a temporary arrangement. Confidential. Married females ask.

Very Single, langur male, searching for harem, will protect against infanticidal males. Estrus or pseudoestrus required.

Kinky S&M Types, hamadryas male wants young inexperienced females. Herding and neck biting my speciality.

Two Wild n' Crazy Guys, saddle-back tamarins ready for one steady gal. Both of us love kids.

CHAPTER 6

Female Choice and Primates

ALTHOUGH PRIMATOLOGISTS' VIEWS of female sexuality have changed, the notion still prevails that females don't enjoy sex as much as males do and that females go about mating at a slow pace, refusing more males than they choose. This prediction of female behavior is supported, as I have shown, not with what females really do, but with evolutionary theory—females *should* be selective, nonsexual. While behaviorists argue in the classroom and laboratory about the impact of female choice, and even whether there is such a thing, and while all the theory piles up in journal articles and all the numbers on mating behavior fill up computer discs, female animals continue to copulate and conceive, regardless of what we scientists think they "ought" to be doing. We are, in fact, miles behind the females themselves in understanding mating behavior. Even now, after at least three decades of long-term primate studies, no one is really sure about the impact of female choice or exactly how female mating behavior affects the evolution of a particular primate species.

We do know that female primates have an influence on who gets to mate and when. We also know that females sometimes prefer some males over others, and there's an inkling that the quality of males, or how they look and act, may make some difference to females. But along the way, some primatologists have also been swayed into thinking that everything a female does is female choice; if she sits next to a male, her action is an example of female choice; if she touches a male, that touch evolved by female choice; if a female mates more often with one male than another, she's choosing the "preferred" male. In other words, some researchers have gone overboard in trying to empower females with their choices while somehow suggesting that the need for choice actually restrains females' sexual activity. How confusing. We keep saying that females must be choosy because they're investing a lot in offspring, and we hope the data will fit nicely into our theory.

More than a hundred years after Darwin outlined female choice as a possible way male characteristics could evolve, and twenty years after Trivers decided that females ought to make choices for their own reproductive benefit, we're still left with the most important questions unanswered. How significant is female choice? Do females have preferences that our human eyes can decipher? If they do have preferences, are they able to execute those preferences as consistent choices that might affect the passing on of genes? What is the evolutionary effect of female desire and choice on the evolution of a species, its mating system, its social system, or the appearance of male traits?

And most important, it turns out, if females sometimes aren't particularity choosy, why not? What do they gain when they fly in the face of evolutionary theory and act promiscuously?

Great Expectations

As I mentioned in Chapter 4, primates, with their powers of discrimination, are reasonable candidates for studies of female choice. We now know that female primates are active participants in the mating game and they have reproductive interests in mind. Sometimes it looks as if they have partner preferences as well. More confusing are species such as chimpanzees and many macaques in which females mate with almost all the males in the group. Their choice for a "lack of choice" leaves theoreticians scratching their heads: females, with their limited reproductive potential and heavy investment in each infant, should know better. Female choice for her own reproductive well being, when it occurs, should reflect the reproductive interests of females.

There are all sorts of things a female might choose for, as I have reviewed in Chapter 4. The problem is, most female primates don't seem to be following the "right" program, nor does each species follow the same program. In this chapter, I evaluate each possible variable that female primates might consider when they choose mates.

A pattern will become clear—the only consistent interest seen among the general primate population is an interest in novelty and variety. Although the possibility of choosing for good genes, good fathers, or good friends remains an option open to female primates, they seem to prefer the unexpected.

A Good Male Primate Is Hard to Find

Animal behaviorists have been so sure that female choice is important, and so convinced that females can evaluate males, that the search for the male qualities that females want has been heated. In every journal, researchers outline why this mammal or that insect should prefer a certain type of male—he's big, old, has high status, has a good territory, will care for the offspring, whatever. The trouble is, it's extremely tricky to substantiate that females actually evaluate males or that they really prefer one type or one quality over another; this difficulty is especially true for primates.

The easiest way to look for a choice for particular characters is to start with the female and write down everything she does. If, for example, she consistently presents to high-ranking males and they respond, she may be choosing for high male rank. Ah, if only it were that simple. The problem is that many variables we humans see as important to mate choice are mixed up and connected—age and rank, size and good genes. It's actually impossible to tease out just one and say that females must be choosing for *that*. In only a few cases are the choices reasonably clear.

The most logical point to look for female mating preference is to consider what a female nonhuman primate must look for. A female monkey first has to find a male of her own species. And so far there aren't any accounts of females mating with males of other species except for a few hybrid cases in which the difference between the males and females is only a few chromosomes, such as the closely related Hamadryas and savannah baboons. But within a particular species, there is some variability among males which might be significant to a female, even among the males in her own group. Obviously, females in monogamous social systems and harem groups have fewer possibilities than do those in groups with many males. But the majority of females have all sorts of options.

Once past the species problem, age, size, and rank are the most obvious differences among males. Age seems an easy variable for females to differentiate. Little males are younger and probably less fertile, and thus might be avoided, but usually they are not. Female primates are known for their mating tolerance. Much of their apparently "promiscuous" behavior is, in fact, mere tolerance of eager little males. These subadult and juvenile males run after females, jump on them, and beg for copulations. Females seem to tolerate them, but most often at the beginning and end of a cycle when pregnancy is less likely. It's amusing to watch a small macque male hang on for dear life to a female's rear end as she continues to eat or groom while he excitedly enjoys his moment of passion. In general, however, female nonhuman primates show a preference for fully mature, and therefore sexually able, males by approaching and presenting more often to them than to younger males, especially during midcycle, when conception is most likely (Small 1989).

The most dramatic demonstration of this preference is seen among wild orangutans, solitary animals who live in harem groups with unique mating patterns. John Mitani, an anthropologist at the University of Michigan, is one of the few people who has studied orangutans in the field. During his many years of observation at the Kutai Reserve in eastern Kalimantan, on Borneo, he witnessed an unusual season of mating (1985b). The three females in his area came into estrus, and he was able to observe their 179 copulations over a sixteen-month period—a record for observed orang sexuality. Orangs live in a social system unlike that of all but one other primate, the small African galago. This system is based on the need to spend all one's time alone; unlike the majority of highly social primates, orangs are basically solitary animals. The mating system is also rather antisocial. Females spend time in small singular patches of forest, and one large adult male typically "owns" these patches. He monitors the females, copulates with them when they're in estrus, and works violently to keep other males away. Subadult young males roam the forest in and out of the resident male's territory, avoiding his attacks and looking for females in estrus. This risky strategy by younger males often pays off—they sometimes get to copulate with estrous females. But the problem for both males is that females don't necessarily cooperate. As Mitani writes, "Achieving intromission with uncooperative females seemed difficult, and males had to grab, bite or slap females before they could copulate. While thrusting, males continued to restrain struggling females by grasping their arms, legs and bodies" (1985b, p. 396). But female orangs do have a kind of preference; copulations with resident males have a fifty-fifty chance of being unpleasant, whereas almost every copulation attempted by a younger male is resisted. Orang females don't exhibit much sexual assertiveness—instead they're completely dominated by males three times their size who fight with one another to gain access. The female's only hope is to struggle against males she may not want, and it appears that she's more averse to younger nonresident males than to the older established male.

A primate female's preference for older ensconced males, when it occurs, is complicated by other male characteristics such as status and size. Older males are usually high status and larger. They may also be, as in the orang case, familiar, because they probably transferred into a group or an area long ago; they may have an established ongoing relationship with a female. In any case, her choice for maturity may be as simple as a choice for sexual expertise and physiologically mature sperm.

Size is also an easy variable that might catch a female's eye. Sue Boinski was the first person to conduct a long-term field study of the tiny but highly gregarious squirrel monkeys of South America (1987). Tiny biege monkeys with black masks, squirrels live in multimale-multifemale groups of about thirty-five animals. The whole group crashes though the forest canopy scaring up insects and searching for fruit, mindless of their squeeking monkey noises that announce their presence to other forest inhabitants. The social life of the squirrel monkey is dominated by female emigration at maturity and the fact that males don't express much of a hierarchy and are rarely aggressivé. Female mating is synchronized, highly seasonal, and estrus appears for only about six days. Males are cued into this short period of heat and they respond by inflating in size by 20 percent, like balloons ready for a parade. They grow larger and larger, brandishing epaulets of fat on their shoulders as females get closer to sexual receptivity. This male blow-up act makes reproductive sense: the largest male is responsible for 70 percent of all matings during the short breeding season. But more important to the female choice story, chubby males don't fight with one another for fertile females, thus male-male competition isn't responsible for their sudden beefy appearance. Instead, female squirrel monkeys actively solicit copulations from the big male. If he is unavailable, receptive females will move on through the forest, calling out to other possible males. But Boibski believes that they prefer the big ones to the small ones, a preference that in turn selects males who are able to pump up at the first sign of female interest.

The search for a connection between certain male genes and what females might like is less clear. No one knows what constitutes "good genes"; even the concept seems rather silly. But because genes are what are at stake here, females may have some way to figure out which male has the best genes to combine with her precious DNA. For example, it's possible that females try to manipulate the genetic makeup of their offspring over time. Because female primates usually produce only one infant at a time, they might be more likely to produce genetically variable offspring only by mating with a different male at each conception, thereby covering their genetic bases over a lifetime (see below). Or, conversely, they might want their brood to be genetically similar in the hope that the siblings express a vested interest in one another. In that case, females might choose males who are related to each other, that is, mate with one brother one year and the next brother the following year (Silk and Boyd 1983). David Smith, a geneticist, and I addressed this possibility when we evaluated the paternity and maternity of rhesus infants born in captivity (Smith and Small 1987). We found that females of the same matriline—that is mothers, sisters, and aunts—did tend to conceive infants with males from the same patriline—that is fathers, sons, and uncles—but we don't know whether females actively sought out those males on purpose. These animals were in captivity, where young males didn't get the chance to move out of the group, and thus there were more related males about than there would have been under normal circumstances.

Female preference for good genes is a given for reproductive success. Female primates mate with males of the same species and most often with older males who are more fertile than youngsters are, but who knows whether they care about the specific genetic makeup of a male? Put more bluntly, we humans don't know what makes a gene a good gene, and neither do other primate females.

Status and Sex

Some argue that rank may be a figment of the primatologists' imagination—we arrange those monkeys into nice neat hierarchies, but do the monkeys themselves perceive a rank order? Dorothy Cheney and Robert Seyfarth, working with the small African vervet monkey, have shown that when rank hierarchies appear, the animals are certainly cognizant of them (Cheney and Seyfarth 1990). The data are simple—when two vervets are together grooming and a third one walks up, the lowest ranking of the three moves away. There's no physical coercion here, just a silent acknowledgment of everyone's place in the hierarchy. If females have a network of dominance relationships mapped out in their heads, they might decide to prefer males of some ranks over others.

We humans always think higher, taller, bigger, more, is better. Thus we also assume that female nonhuman primates should like high-ranking males more than low-ranking males. But what does high rank in her partner really gain a female monkey? In most cases, female nonhuman primates don't need much from males. Females forage on their own, and even when high-ranking males sequester food, as in chimpanzee hunts in which males grab most of the meat for themselves, males rarely share that food with females. What a female might gain from high-ranking males, in the immediate sense, is protection from others, including outsiders that are threatening the group as well as those on the inside that present a more intimate menace. But in the bargain, she might get harmed herself. Male primates are notably more aggressive than females, and a female close to a male could get in the way of two contentious males. In addition, any protection or increase in personal status a female might gain by mating with a high-ranking male is transitory—as soon as the copulation is over, she's back to her original position and without protection. Females might gain something else from high-ranking males—good genes. High-ranking males are often larger, more personally assertive, and they might be older. Any female would be impressed with this vigor, and if rank order is at all influenced by the inherited components of behavior and personality, the female could pass those presumably excellent winning genes on to her sons and daughters. Mating with high-rank males seems a good idea.

There's some evidence that, in a few species of nonhuman primates, females really do prefer high-ranking males. I found that observers were convinced that the females of nine species of primates favor high-status males, and these status seekers range from the tiny African galago to two species of baboons (Small 1989). But nine species really aren't very many when we consider that there are two hundred species of primates and that we have decent reproductive research on at least twenty-four. In addition, the individual studies, on closer look, don't appear that clear at all. Females don't always like high rankers, and they often mate with males from other rank orders as well. The issue of female preference for male rank is as flexible as the females themselves.

For example, Charles Janson, an ecologist at the State University of New York at Stoneybook, followed capuchin monkeys in the Manu forest of Peru (1984). This must have been fun—capuchins are better known as organ grinder's monkeys, full of personality and mischief. But Janson's description of the females is nothing close to that perky monkey in a hat and vest perched on the arm of a trainer. Instead, his females were often frustrated sexual dynamos. Capuchins live in small groups that include about four adult females. There's no real breeding season, but they do exhibit a breeding peak, when many of the females might be in estrus simultaneously. It would be better if they weren't. Each female becomes a whistling, whining mating machine, and the object of attention is the alpha, or highest-ranking, male. A female in estrus will run up to the male, grimace at him, slap him lightly, and run off in a game of sexual tag. Her ardor is so strong that she may not bother to eat during this period of sexual excitement; she just continually pesters the chief male. He, in turn, isn't necessarily up to the challenge, especially if there's more than one female after him at a time. The alpha male, on average, mates only once a day. What's a female to do? In response to his rejection, the female eventually gives up and moves on to lower-ranking males. Thus her preference for his rank is diffused by her inability to gain access to the male she wants. In essence, her preference and choice for the highest-ranking male translates into nothing: these females are highly promiscuous. One female followed by Janson mated with four different males in ten minutes. It seems that even when rank may be at issue, a female must above all be concerned with copulation in the first place.

The problem seems to be the difference between what females might want and what they actually get, and male desires keep cluttering up the female system. This point has been well illustrated by Ann Keddy with captive vervet monkeys (1986). Vervets normally live in rather large groups with several males and females. These females aren't necessarily dominated by males. In fact, female-female coalitions can easily make an aggressive male back away, and female power in numbers is a driving social force in this species (Cheney and Seyfarth 1990). Females have no external signs of estrus (Andelman 1987), and the breeding season is very tight, only two months out of each year. Keddy conducted a series of experiments on captive vervets to determine whether females made mate choices based on male-dominance status, and she designed the study so that females were entirely free to express those preferences—she wiped out competition from other females and other males to see what happened. When Keddy placed only a pair of potential mates together, females made approaches to high-ranking males more often than they did toward low-ranking males, and females rarely refused high rankers. At the same time, the females tried to resist low-ranking males when left alone with them, but only high-ranking females could stop the low-ranking males from mounting; low-ranking females needed their female allies to stop a mating they didn't want. And even more interesting, when the groups were reunited, low-ranking females, who have little power in a social group, quit approaching their preferred males. They were apparently stopped by the fear of competition from their higher-ranking female troop mates. Sandra Andelman, who watched vervets under natural conditions, points out that vervets usually reject male approaches as much as possible (1987). Keddy's work suggests that the vervet situation is confused by the interaction of male and female power. Females may want high-ranking males, and they try to refuse others, sometimes unsuccessfully. This situation makes for endless sexual squabbles in this species. The result is that, over the long term, male rank has no real effect on who gets to mate (Cheney et al. 1988). Females may prefer high rankers, but they can't seem to have them exclusively, and females must constantly protect their flanks.

Several studies have demonstrated a positive association between male rank and high reproductive success (Curie-Cohen et al. 1983, deRuiter et al. 1992, Martin and Dixson 1992, Smith 1982). David Smith and I used paternity identification data of the captive rhesus at the Davis Primate Center to look at the relationship between male and female rank: who exactly conceives with whom, we asked. We found that high-ranking males produced the most infants, and they did so with high- and low-ranking females alike. Low-ranking males conceived more infants than we might have thought, but they mostly got to father babies of low-ranking female peers (Small and Smith 1982a). But no one is sure how much females influenced this connection. High-ranking males almost always have silent privilege, and without any show of threat they gain females. And when there's an altercation, the male of greater status wins the females. At the same time, even if females want high-ranking males, they often are unable to get them. But does this really matter? No one is sure what females really gain by mating with high-ranking males rather than with their subordinate troop mates, and females *always* turn to low rankers when high rankers are unavailable.

Male rank, then, appears of equivocal importance in the scheme of female mating strategies among nonhuman primates, and surprisingly so. In our own society, high-status males are important in the societal sense, and there may be a relationship between human male status and greater reproductive success (see Chapter 7). Status, hierarchies, and rank are certainly part of our primate heritage, but unlike ourselves, nonhuman female primates seem to focus less on status than we do. A fancy car or expensive dinner out probably won't do a male ape much good when looking for a mate.

It Takes Someone Special to Be a Daddy

One of the major features of the order Primates is the months and years of infant dependency (see Figure 6). Babies are most often carried around, if not by the mother, then by another. Some strepsirhine primates, such as the galagos, park their infants in branches and visit them at regular intervals. But most primates don't have that luxury. The infant needs to be fed too often, and predators lurk about. These aren't babies who can jump up and run away at the first scary noise—they have to be lifted to safety. The long years that infants hang around primate parents foster an atmosphere of extensive learning, and learning proves important in this ever-changing world. The trade-off in this evolutionary deal is that primate babies are born relatively unfinished, and their brains continue to grow when they leave the womb. One reason primates are smart, the reasoning goes, is this long period of being less than grownup. And there seems to be a direct correlation among the size of the species, the size of the adult brain, and the time spent dependent as a youngster. For example, lemur infants have small body size and smallish brains, and they're dependent only for a few months. Chimpanzee infants, in contrast, are rather large in comparison, they have huge brains, and their period of youthful dependency is at least six years.

One might expect that it could take more than one parent to bring up the dependent vulnerable primate offspring. If this were true, it would behoove a female primate who needs infant care to choose a good father. This line of thinking is certainly familiar. We humans have the longest period of infant dependency, and very weak infants. They must be carried, fed and protected; they can't even hold up their heads for several months. Males as parental partners have probably been important in the human lineage for millions of years (see Chapter 7), and one concern a human female might have in deciding on a mate is his fathering potential. When fathering is important, other female primates perhaps should also look for good fathers. Do they?

There's one logistical problem, and this is as true for human females as it is for monkeys—how can a well-intended mother evaluate a "good father"? Or, put more bluntly, how can you know if a male would be a good father until you see him *act* as a father? And do males have strategies to convince females of their daddy factor?

Ann Keddy again used vervet monkeys in captivity to test if males would act differently toward infants when mothers were about (Keddy, Seyfarth, and Raleigh 1989). Male vervets don't take on any infant care, but Keddy thought that they might want to impress females with some semblance of paternal behavior. Keddy tried three experiments using high- and low-ranking males and pairs of mothers and infants. The test housing consisted of two separate chambers. On one side she released a male and an infant, while the infant's mother occupied the other side. Sometimes the chambers were separated by an opaque partition and neither side could see what the other was doing. Sometimes a one-way mirror allowed the mother to see the male-infant pair although they couldn't see her, and sometimes a transparent shield separated the mother from the pair so that everybody could watch everybody else. Keddy found that when the male could see the mother he acted quite differently, especially if he was a low-ranking male: males were much more friendly toward infants in this situation. The one-way mirror created a real problem for males who thought they weren't being watched. When the mother rejoined the pair, she was highly aggressive toward a male who had been unkind to her infant, threatening and attacking the unsuspecting offender. This study suggests not only that females are attending to the behavior of males but also that males actually change their "paternal" behavior to impress females. To what extent this behavior alters male-female relationships, or influences female choice or preference at mating time, we don't really know.

From an evolutionary standpoint, a male should be a good father only to his own offspring. Why invest in some infant that doesn't even have any of your genes? In most primate groups, females mate with more than one male. When paternity is uncertain, we can't expect much from males, and as predicted, we don't see much fathering beyond mere tolerance or the occasional play bout (Taub 1984). Some male macaques carry infants around, but they really do it for their own purposes. Male Barbarys, for example, pick up infants and carry them to more dominant males. The infant acts as a "passport" or "buffer" in a tense situation. There's no pretense that male "care" has anything to do with previous or later mating access to females (Deag and Crook 1971, Small 1990d, Taub 1980b). Males are "using" infants, not fathering them. The only evidence of a male's interacting with his possible infant in a species in which females mate with more than one male comes from work on chacma baboons in Botswana. Curt Busse noticed that when extratroop males came into the area of the group he was watching, resident males often picked up infants, as if to protect them from the possibly infanticidal interlopers (Busse and Hamilton 1981). Counting back from the day of the infant's birth, Busse discovered that the protecting male might have been the infant's real father: he'd consorted with the female at the right time. The problem here, of course, is how did the male "know" which infant was his? It's possible that baboons keep track of mating that occurred six months previously, but more likely the "fathering" male has some sort of ongoing relationship with the infant and its mother, and the conceptive consortship, as well as the fathering, is part of a general pattern of association. More often studies have shown that there's no way males in multimale groups could know who fathers which infants (Berenstain, Rodman, and Smith 1981, Curie-Cohen et al. 1983, Stern and Smith 1984), and females don't seem to choose males based on any paternal behavior. Part of the complication may be that females rarely have infants who share the same father (Berenstain, Rodman, and Smith 1981)—more to the point, females seem to prefer having infants with different fathers, and thus paternal behavior is of little consequence.

The issue of good fathering might be more important to females who operate in a somewhat monogamous breeding system. For these few species, choice for a mate can often be for life, or at least for many years. Interestingly enough, paternal care has not evolved among all the monogamous primates. For example, Indri males, the large lemurs of Madagascar, are monogamous, but males don't act like good fathers. Gibbon males also don't pay much attention to infants, although Ryne Palombit's work shows that female and male gibbons sometimes sneak out of their monogamous circle and mate with others. Perhaps the male gibbon has not evolved parenting skills because he is not strictly monogamous, and "his" infant may be some other male's.

For many monogamous species, however, paternal behavior is crucial to infants. Soon after birth, South American marmoset and tamarin infants transfer to the backs of fathers. The males aren't just good fathers: they're the primary caretakers. It's amusing to stand at the zoo, in front of the tamarin cage, and listen to other visitors extol the virtues of the mommy carrying the baby. In fact, if they looked a little closer, they'd see that the caretaker was a male, and the infant hops back to the real mother when it needs to nurse. Given the importance of fathering to females in these groups, one might expect females to pay close attention to males' parenting skills. In addition, males ought somehow to show off these skills. A group of cotton-top tamarins housed in Sterling Scotland has shown just this relationship between mated pairs (Price 1991). After tamarins give birth, they come into estrus again only two to four weeks later. At the same time, the new infant is still clinging to Dad or Mom. It appears that males use this birth/estrus overlap to demonstrate their dedication to fathering, and females pay attention. When a male mounts a female, she's less likely to reject him if he's carrying the current infant. Males also attempt these mounts more often with infants on their backs. This is clearly a direct stimulus that tells females about fathering skills, and females seem to reinforce these skills by being willing to mate again with the same male.

Other than this study, there's no real evidence that female primates choose males for fathering skills. Monogamous primates may do so, for their choices are imperative, but no one has observed the initial mate selection by a female for a male when a monogamous pair is established. So far, therefore, choice for good fathers sounds like a good idea, but one for which we have only weak evidence.

A Friend in Need

Intuitively, we believe that female primates "ought" to mate with males they know. Mating is often such a dangerous liaison for females because males are none too nice (Smuts 1992, Smuts and Smuts in press). They herd females, try to keep them away from other males, and fights break out when estrous females are about. Most human females also think that it would pay to choose the "nicest" male and stick with him. But again, our human glasses may be in the way of seeing just how little importance female nonhuman primates attach to knowing a male before they mate with him.

The idea that friendship and familiarity should be important to females comes from extensive work on baboons. For years, baboon researchers have reported on what they call "special relationships" between particular males and females (Ransom 1981, Seyfarth 1978, Strum 1975). In 1985, Barbara Smuts published *Sex and Friendship in Baboons*, which started a virtual friendship craze in primatology. The photograph on the cover of the book is enough to make you sigh— a burly brown baboon male and his petite female lean against each other in friendship. Smuts discovered that most baboon females have at least two special male friends. Females who aren't cycling tend to stay close and groom with these friends more often than with other males, although these females also regularly interact with other males. Males, in turn, provide these females and their infants with protection when troop members start fights.

But the issue of friendship and its importance to female choice is not as clear as it sounds. First of all, neither Smuts nor anyone else has shown that friendship leads to preferred access during mating. Friendships actually start up during the time the female is cycling and extend into the female's pregnancy. After an infant, not necessarily the offspring of one of the friends, is born, the male friend often forms an attachment to the infant which might last years. But does friendship actually allow a male more access to a female during this or her next fertile period? There's no clear evidence that friendship actually helps a male improve his reproductive success. Males do have consortships with female friends, but this arrangement isn't exclusive for either the male or female baboon. In fact, many consort partnerships in Smuts's study, especially those with older males, weren't among friends at all. And in other baboon studies, females often prefer to mate with nonfriends when not engaged in a specific consort (Rasmussen 1983, Seyfarth 1978). More important, no one has been able to determine paternity in these groups to see whether or not males actually father the infants of their female friends. And while friendship might have other benefits that don't directly affect conception, such as male protection for females or their infants, no one has shown a consistent and impressive relationship between friendship and actual preferred choice during estrus by females.

Other studies on macaques have shown that friendship might be more of a reproductive deterrent than an attractant. Female Japanese macaques also have sort-of friends. These relationships go on for years, but the longer the relationship the less likely a male will be allowed to copulate (Baxter and Fedigan 1979, Enomoto 1978, Takahata 1982a, 1982b). And it's the female who calls the shots in the friendship when copulation is at hand: the male continues to attempt a mount, but she refuses. Work on rhesus macaques shows that males and females sometimes have special relationships too, but these friendships have nothing to do with mating (Chapais 1986). Apparently rhesus and other macaque males stay near high-ranking females just to stand in their spotlight and let some of the female rank privileges rub off (Gouzoules 1980, and personal observation). It's possible, researchers suggest, that Smuts's view of baboons friendship is colored by the short duration of her study, only two years. Perhaps in time she too would see these friendships dissolve as females reach out to new, immigrant, and more unknown males.

The point is, the relationship between an established friendship and mutual attraction when females are fertile is equivocal (Berkovitch 1991). There may be some effect, but it's never strong over time, nor is it universal for all primate populations. More important to the issue of female choice, although friendships between females and particular males occur, they don't necessarily affect the way females select their mating partners.

And Now for Something Completely Different

Contrary to most expectations about females and what they "should" want, female primates have a strong penchant for the unusual, the novel, the unfamiliar male. This preference is odd, given that females place themselves at great risk when they slide away from group males and seek out the unknown, but female nonhuman (and perhaps human) primates seem to be pulled in that direction. The generic monkey, the rhesus macaque, is the most demonstrative about her penchant for the unfamiliar.

In the 1940s, Carpenter alludes to more than a hint of female promiscuity—he writes about female rhesus who sported several vaginal plugs of male ejaculate and participated in multiple consortships during estrus. "These observations indicate, among other things, the enormous excess of spermatic fluids delivered to females beyond the amount necessary for fertilization" (1942, p. 135). He also refers to a female's "sexual hunger," and decided that a female's capacity for mating exceeds that of any male. "A single estrous female," he writes, "may satiate, entirely or in part, several sexually vigorous males; although when a period of such intense sexual activity terminates, the female may show extreme fatigue bordering on exhaustion" (1942, p. 141). This certainly isn't the picture of sexually uninterested females that behaviorists in the 1940s might have expected.

For almost fifty years, no one asked why rhesus females were participating in all these "excess" matings, not even Carpenter. Although he documents females with a succession of males during estrus, mentions intense female aggression and improved social power, he repeatedly emphasizes that females operate only within a sphere of male domination. It would take the feminist revolution and the liberation of female human roles in Western society for researchers, male and female, to reevaluate rhesus female mating from the females' perspective. The answers would be quite surprising, and perhaps shocking to Carpenter.

Most recently, Joseph Manson of the University of Michigan, armed with the contemporary concepts of sexual selection theory and mate choice, directed his Ph.D. dissertation to the issue of female choice among rhesus (1991, Manson 1992). It seems odd to realize that after decades of intense observation of rhesus females—probably including more observation hours than for any other primate species—we still hadn't looked specifically for the impact of female choice. Manson's study is unique because he didn't assume that choice necessarily occurs, and he set out not to prove that females make choices but to find out how one could determine if they do. For two breeding seasons, Manson and his coworkers watched female rhesus in estrus at Cayo Santiago. They wrote down everything the females did in two-hour blocks, following them through the forest, out on the cliffs, and down to the beach. He knew that teasing out female choice would be difficult. Is a female choosing to participate in a copulation or is she being coerced by subtle cues given by a male, cues we humans don't perceive? Manson decided to concentrate on four behaviors: proximity (which in animal-behavior language means whom the female is sitting close to), sexual presents, sexual refusals, and what happens when a female is with one male but is harassed by another male.

He discovered that female rhesus exert an influence on the mating system in several ways. First, females refuse to cooperate in many of the sexual solicitations made by males. When a rhesus male is interested in a female, he grabs her with two hands on the rear. If she's sitting down, he pushes her up. But the female is actually in charge here. She can refuse to stand up, or she can walk away from him. The females in Manson's groups were not always cooperative—they stayed seated or moved away in a third of the cases. They weren't choosing *for* particular males, but they were certainly choosing *against* certain males. Manson also discovered that female rhesus are responsible for maintaining proximity to males. Most dramatically, almost every hour high-ranking males try to break up consortships of low-ranking males with a female. But after an attack by the high ranker on the consorting pair, the female will reestablish the consortship, and not with the attacker but with the low-ranking male of her initial pact. Bully rhesus males are thus thwarted by the female decision to continue with a relationship, usually with a lower-ranking male.

But the more revealing conclusion from Manson's data is the overall pattern of female mating behavior. Female rhesus on Cayo, it appears, like a variety of males. By virtue of their status, high-ranking males get to copulate more often than do low-ranking males, but females are instrumental in seeing that low ranking males also get plenty of opportunities. The same sort of behavior has also been observed for wild rhesus in India (Lindburg 1983). Even in small groups, in which females have less opportunity to sneak off and mate with peripheral males, they still manage. Whenever Donald Lindburg discovered that one of his estrous rhesus females was temporarily missing from his group in India, invariably a low-ranking male was also missing (1983). And in one case, when a male transferred into a group, his copulation rate was higher than that of any other male and he was constantly followed and solicited by females.

Manson thinks that females have to be more assertive when they seek low-ranking males. Lindburg agrees when he says that the highest-ranking male in Indian groups never receives sexual invitations from females, but other males do. The high rankers, the males of privilege, will always get what they want; females can just hang around and high-ranking males will come to them. In a true sign of female preference and selection, however, female rhesus have to go after lower-ranking males wary of attacks from other males. As a result, female sexual assertiveness has an effect on male copulation rate by evening out what would be a skew toward high-ranking males' getting all the copulations. (See Figure 7.)

Manson's take on this situation is interesting. He saw females going to a lot of trouble, putting themselves at risk from aggressive high-ranking males, just to be with low rankers. Why would they do this? He feels that rhesus females have an urge for a mixture of males, and this is the curious, and significant, result here. When females sit next to low-ranking males, they're at risk of attack, but the stalwart females continued to stake out these low rankers and put up with abuse from other males. This behavior is perhaps the clearest evidence of some sort of female preference, and the preference seems to be for variety.

As female rhesus make moves to include low-ranking and peripheral males, the consequences of their behavior reach beyond the possibility of mating with a variety of males. Their behavior, may in fact be the driving force behind the evolution of the rhesus social system.

John Berard has been working on the Cayo Santiago rhesus continuously for almost eight years. Recently he tied together a number of his studies on both males and females. Like Manson, he watched females, but he and several research assistants followed females all day long, not for just a few hours. At the same time, he was watching males as they transferred in and out of groups. With the unique situation at Cayo, where all the groups are known and where males walk from one identified and censused group into another, researchers track male movements and they know the results of emigration and immigration. Berard found that males who leave a group tend to be males with pitiful breeding success. They don't do very well their first year in a new group either, but their rate of copulation increases the second year of tenure. Berard believes that loser males are not forced out of their natal group but that they leave under their own volition to seek matings elsewhere. But more interesting, Berard has a hunch that female preference is driving male mobility. Like Manson, he saw the female predilection for low-ranking males who had immigrated into the group a few years back. DNA paternity identification of these animals shows that the "new" males are actually fathering the most infants (Berard et al. ms.). Each year, females prefer newcomers, and thus males who move in from other groups have higher reproductive success. Resident males actually experience a downward slide in the number of infants they father each year. Berard suggests that female choice for outsiders is driving the evolution of the male transfer system among rhesus. Why females follow this path is unclear, but the best explanation is that females are wise to inbreeding—they look for new males who have genes different from their own.

The female choice–male transfer system has also been suggested by Michael Huffman for Japanese macaques (Huffman 1992). These female Japanese macaques reject males almost half the time, and they even reject high-ranking, older males. Some of these unwanted males are even their "friends," but instead of friendship leading to access, familiarity breeds contempt; after a few years of friendship, females aren't interested in copulating with familiar males. Huffman, too, suggests that when females reject males they inspire intertroop movement. Other macaque watchers have gone even further. Yamagiwa (1985) suggests that when females establish consortships with peripheral males demographic changes occur in the troop. In small groups, for example, there may not be enough males to go around when breeding season is in full tilt. Females then seek these outsiders, and in the process the group may split apart, forming two distinct groups.

The same phenomenon is happening among troops of savannah baboons. Females who aren't yet in consortships are intrigued by extratroop males, and these males often transfer into their group. Unattached females present to alien males at a higher rate than they present to resident males, and the unfamiliar males receive all sorts of interest from different females (Berkovitch 1991). In fact an outside male's transfer into a baboon group is facilitated, and sometimes only accomplished, when he attaches himself to a female.

And it's not just among the higher primates that the female preference for outside males has an important effect on the social system. In two studies of ringtailed lemurs, one on animals in captivity and one in the wild, females consistently seek out nonresident males (Pereira and Weiss 1991, Sauther 1991). Ringtail females have an extremely short estrus—only six to twenty-four hours—and they synchronize this period of heat within less than a month of one another. Females are also powerful and dominant to males at feeding sites, and usually the same size or larger than males. During breeding season, when males spend all their time scrambling for females, the males appear harried and thin. Although females often prefer the highest-ranking males in their group, they are just as likely to head right for males who have recently tranferred or those in neighboring groups. Ringtail males transfer among groups on a regular yearly basis, a round-robin of male membership for each group, which stays fluid in this way.

How females implement this orientation toward unfamiliar but inherently interesting males varies. They might just sneak into the bushes with a newcomer while resident males are occupied elsewhere, or lure new males closer and closer. Smuts tells the story of one young female baboon who lured a male away from his group and into her own troop by repeatedly presenting her swollen rump in his direction (1985). The male was so distracted by this display of fertility that he forgot that entry into a new troop was a risky adventure.

Captive lion-tailed macaques at the San Diego Zoo have also been providing intriguing information about how females might be active temptresses. Over a six-year period, Lindburg and his colleagues noticed that female lion-tails emit a copulation call even when they aren't copulating (1990). The call is clearly a "come hither" signal: a female stares at a male, rubs her genitals, and sometimes presents. The call occurs only during the first half of the female's cycle, and Lindburg speculates that it indicates a readiness to mate. But what caught the eye of lion-tailed observers is the target of this invitation. Females in estrus direct the call most often to males outside their group. They orient toward males housed in single cages outside their normal group, and when the researchers placed an unknown "stimulus" male in a cage within view, the females oriented toward him. The only time females did not favor extratroop males was when their breeding male had been replaced rather recently. Lindburg believes that females direct calls toward foreign males in order to mate with unfamiliar partners. Although intertroop copulations have not been seen in the field, there's some indication that such an exchange is likely. When female lion-tails come into estrus during the summer months, the number of intertroop encounters increases, and resident males try to herd females and chase them away from males in other troops (Kumar and Kurup 1985a). In other words, female lion-tails may be calling out to these unfamiliar males and bringing them within striking distance, altering daily travel patterns and possibly changing the social structure of a group.

This interest in novelty has been clearly documented for twelve species of primates (Small 1989). But more striking is the regularity with which females choose these males. Female chimps leave their natal group, copulate with neighboring males, and then return; patas monkey females roam the savannah looking for males other than their harem leader; squirrel monkey females seek extratroop males; monogamous gibbons also look to neighbors. In fact, the search for the unfamiliar is documented as a female preference more often than is any other characteristic our human eyes can perceive. Perhaps these females are concerned with inbreeding, and seeking new males is an easy way to ensure a mixture of genes for the offspring. Females, rather than males, should be the most worried about inbreeding because they have the largest investment in each conception (Huffman 1993). Their interest in novelty, therefore, might be a primitive urge to find mates lurking at the periphery of their group with genes different, but not completely different, from their own. Suzanne Ripley has suggested that avoiding inbreeding and gaining genetic diversity would be especially important in multimale-multifemale groups in which an ability to adapt in a changing world is paramount (1980). The evidence that this might be true comes from the genetic makeup of the animals themselves. Rhesus monkeys, perhaps the most evolutionarily successful primate after humans, are also highly genetically diverse (Melnick, Pearl, and Richard 1984, Ober et al. 1984). Although other authors have attributed this genetic diversity solely to male migration, it may be that females are more responsible in that they prefer to mate with outside males and place a forward spin on male transfer (Lindburg 1969, L. M. Wolfe 1986).

In this sense, female choice is not merely a matter of females' choosing some males over others and making an impact on the genetic component of future generations, especially of their own offspring. For rhesus, Japanese macaques, and baboons, and perhaps for other nonhuman primates, the entire social system may be driven by what females prefer. The power of female choice, in this instance, is not just for mating—it exerts a potent evolutionary force on the entire social system.

Promiscuity: A Lack of Choice?

"Promiscuity" is most often applied to females, both human and nonhuman. It means indiscriminate sexual behavior, casual, lacking choice. But there's also a pejorative hint to the word, probably *because* it refers to female behavior. I think that humans are disturbed when they see "promiscuous" females because U.S culture still frowns on female sexuality in general. But more important for the primate story, evolutionary biologists have trouble figuring out why some females bother to mate with more than one male, or so often with one male, when a single well-timed insemination would do. Many primates don't seem to be doing what evolution says they "should" do.

It seems that primate females, when the bonds of male domination are loosened, mate with more than one male. Barbary females float from male to male, copulating in random fashion with every male they come across (Kuester and Paul 1984, Small 1990b, Taub 1980a); Japanese macaque females have as many as nine different mating partners during each estrus (Huffman 1991b); South American howler moneys also mate with many males in their troop and pay little attention to rank (Jones 1985); chimpanzee females mate with several males if they choose not to take off on a mating safari with one male (Goodall 1986, Tutin 1979). One wild rhesus female, observed in India by Donald Lindburg, shifted among four males of her group in the span of two hours and only the highest-ranking male turned her down (1983). Woolly spider monkeys who grace the high canopy of remnant forests in Brazil are also highly promiscuous. The anthropologist Kathryn Milton knew a female was in estrus when at least seven males suddenly appeared in her foraging area (1985). Milton hypothesizes that woolly spider females signal their suitors by traveling widely in the area, marking branches with pheromone-filled urine and twittering in the direction of males. The love-hungry males hang close to females and copulate whenever they get a chance. Milton counted as many as four different males copulating eleven times in a single day with the same female; that's about once an hour for each. When the males lost interest and began to leave her area, the female chased after them, calling and soliciting. Even when male and female woolly spider monkeys live in heterosexual groups, as they do in Karen Strier's field site in Brazil, females mate with every male around (1990, 1992).

Baboon females, who are actively herded by consort males, sneak off with other males or try to signal that they want out of a current consort (Bachmann and Kummer 1980). Baboon females also increase the number of consort partners on the days surrounding ovulation (Strum 1982). Nonconsort sexuality among chacma baboons is so wild that one researcher called it a "roving appetitive behavior" and said that it is common for a female chacma baboon to present and mate with as many as three males in three minutes (Saayman 1970). These females try to get the attention of males at twenty feet with a special sexual face. The face is stretched back, the ears go flat, the eyebrows are raised to expose white eyelids, and the female lip-smacks in his direction.

Not all nonhuman primate females are "promiscuous." As with all behavioral patterns, promiscuity is a word that describes only one end of a continuum. On that end lie Barbary and other macaques. On the other end are, perhaps, gibbons, who almost never copulate at all. The point is, most females mate with more than one male or give some indication that they'd like to mate with more than one if given the chance. Even more important to the female primate story, nonhuman primates are showing us that females are definitely not passive sexual partners.

The Promiscuity of Our Closest Relatives

We're more closely related to the apes than to prosimians and monkeys, and thus their mating behavior should be of special interest. We're even more closely related to chimpanzees and bonobos, sharing about 98 percent of our genetic material. Interestingly enough, the sexual behavior of these two species is markedly different, and it may be a toss-up which one we're most like. In any case, chimpanzees and bonobos merit special consideration here.

Male chimpanzees are usually unkind to females, often threatening and attacking them for no apparent reason. Males are very tolerant of one another, however, and most researchers believe that this tolerance reflects a genetic relationship; the males are highly related, perhaps even brothers, and thus don't bother to compete with one another. But male chimps have huge testicles and high sperm counts, and thus they must incur some competition among themselves for reproductive success, even if it's inside the female reproductive tract. Although the males are tightly bonded, females spend much of their time alone or with their offspring. When a female comes into estrus, with her huge pink swellings, she often travels with the band of males. At the beginning of her cycle, she'll mate with just about every male (Goodall 1986, Hasegawa and Hiraiwa-Hasegawa 1990, Tutin 1979, Tutin and McGinnis 1981), but near her midcycle, a high-ranking male might try to keep other males away. He will try to lure the female off on a safari where only he has access to her fertility. What's striking about chimpanzee sexual behavior, and the female role in sexual interaction, is the variety. In a group setting, a female in estrus is the belle of the ball. Each male spreads his legs and waves his penis in her direction as an invitation. It's up to her to decide. Japanese researchers estimate that chimpanzee females copulate 135 times for every conception—the average female chimp has three cycles, lasting about 12.5 days, during which they copulate about 3.6 times a day (Hasegawa and Hiraiwa-Hasegawa 1990)—and no female mates with just one male. Females rarely refuse an opportunity to mate. More telling is that most conceptions occur during matings in a group setting, not on those safaris away from the group (Goodall 1986). The exclusive mating, in which a male and female pair up, also occurs within a group setting, but it's difficult to tell if these more exclusive matings reflect female choice. When a male tries to lure a female away for exclusive access, she can also refuse, but uncooperative females are often brutally attacked by their would-be suitors. In addition, a long history of male intimidation may play a major role in the lack of female refusals under any circumstances (Goodall 1986, Smuts 1992, Smuts and Smuts in press). Thus no one is sure how free female choice is in chimpanzee society. But it does appear that females make some choices among males, and selection seems to be based on what Jane Goodall calls "personality." Nicer males do best.

The most flamboyant example of female primate promiscuity comes from our other closest relative, the bonobo. Disturbing to some is the fact that we're as closely related to bonobos as we are to chimps, and bonobos are the most sexual animals on earth.

I was first introduced to female bonobos as promiscuous primates by the photographs in Frans de Waal's report in *National Geographic Research* (de Waal 1987). A veteran of years of watching the sexual behavior of monkeys, I was still amazed. There were face-to-face copulations and the silly happy grins of females and males (see Figure 8). It looks so human—actually better, and even more fun, than human sex. In December 1991, I attended a national conference on chimpanzees and bonobos and saw on video what no still picture could really explain—bonobos of all sexes and ages playing sexually with one another. These films, by Amy Parish and Frans de Waal, silenced a room of three hundred primatologists and journalists. Watching this ape sex-play is like watching humans at their most extreme and perverse. Bonobos have sex more often and in more combinations than most people of any culture, and most of the time bonobo sex has nothing to do with reproduction (Small 1992c). Males mount females and females sometimes mount them back; females rub on other females just for fun; males stand rump to rump and press their scrotal areas together. Even juveniles participate by rubbing their genital areas with adults, although males never actually insert their penes into juvenile females. Very young animals also have sex with one another—little males suck on each others' penes or French kiss with their friends. When two animals initiate sex, others are free to join in by poking their fingers and toes into the moving parts, or they may be the next in line. More striking is the often-used face-to-face copulation between males and females, rarely seen any animal other than humans. Although males most often mount females from behind, females prefer face-to-face. Scientists assume that the female preference is dictated by bonobo anatomy; the female's enlarged clitoris and sexual swelling are oriented far forward. Females presumably prefer face-to-face contact because it feels better.

Sexual behavior of female bonobos is facilitated by basic biology. Unlike most other primates, human females don't have any external signs of fertility and they can copulate at all phases of the menstrual cycle. Many other primate females sport large pink swellings on their hind ends which signal fertility, and they are interested in sex only during prescribed, and often very short, intervals. With bonobos, the swelling never seems to wax and wane very much, and females are always ready for sex: their pattern of continuous sexuality is more like humans'. A female bonobo copulates during any time of the estrous cycle; she copulates whether or not she has a sexual swelling, during pregnancy, and within a year after an infant is born even though she is still nursing. And she utilizes this boundless sexuality in all her relationships. Because sex is detached from reproduction, it can be used to fool males and gain food or equal status. By wielding their sexuality, females have gained easy access to important resources and liberation from male authority. One might ask, what then happened to female sexual liberation after humans split off from the common ancestor? And why did bonobos follow the more egalitarian road while chimpanzee and prehuman females walked right into male domination? Our evolutionary history, covered in the next chapter, addresses these issues in more detail, but we must accept that our common ancestor may have had touches of the sexuality of present-day bonobos.

Why Promiscuity, Novelty, and Variation?

The most striking feature of female primate sexuality is the consistent orientation toward novelty and variation, sometimes to the point of promiscuity. The behavior of these females can be interpreted as wanton, random, or purposeless, but that interpretation would underestimate the mating strategy of promiscuity. An obvious question follows: what advantages do females gain from mating with more than one male, a variety of males, or those risky outsiders?

Perhaps the female interest in multiple males is just so much baggage (Halliday and Arnold 1987). Males have been selected to chase after many partners, and it might be that females are still riding on the sexual coattails of males on their way to more discriminatory behavior. But this explanation doesn't carry much weight given the evidence that females seem to have their own brand of promiscuity, and it's seen in so many species.

There may, instead, be social functions to mating with many and novel males which have little to do with reproduction and much to do with staying alive as a female primate. Most primates are social animals, and presumably mating is part of the fabric of primate sociality. When female nonhuman primates mate with multiple males, they extend the normal social interactions of grooming, sitting together, and eating with others to sexual behavior. Females who mate with several males rather than with one or two may be trying to broaden their horizons and their social network. But it doesn't seem that a few copulations would make much of a foundation for significant attachments. There just isn't good evidence that sexual behavior carries over into other social realms for most female primates, except perhaps for bonobos.

Sarah Blaffer Hrdy (1981a) and Tim Halliday (Halliday and Arnold 1987) have outlined several other social possibilities that might account for female wanton behavior. First, females who mate with many males can better evaluate a series of males and choose the "best." Sexual swellings signal males and may elicit competition with obvious winners for females to choose among. More dramatically, female philandering may save the female and her next offspring from disasters. This suggestion comes from Hrdy's years studying Indian langurs; she records that males often kill infants during troop takeovers. Infanticidal males are found throughout the primate order (Hausfater and Blaffer Hrdy 1984), and it makes sense that females would solicit males, and mimic sexual receptivity, to save their infants. But not all primate species experience infanticide, and females are usually mating with more than one male in less dangerous circumstances. More to the point, promiscuity most often takes the form of a female's moving from male to male or agreeing to mate with most males who approach. In these cases, Baffler Hrdy and others are inclined to see a connection between female mating and later paternal care (Stacey 1982). A female might manipulate the system enough to stay with one bonded male who will be a good father, but then sneak off and conceive with another male with better genes. Or a female could confuse paternity among several males and thereby gain care, or tolerance, from them all once the infant is born. But so far there is absolutely no evidence that promiscuity gains any infant care from males. Females could hypothetically gain things other than parental care. Unfortunately the only species in which a trade for sex makes much sense is among humans. Other primate females do perfectly well in the food and shelter department without selling their sexuality for favors.

These are some possible social reasons for less-than-choosy sexual behavior, and they may have some bearing under special circumstances. But the substantial evidence for female promiscuity and the search for novelty can't usually be explained by social reasons alone. If evolution is responsible for this pattern seen in many primates, there's probably a direct reproductive effect at work. A physiological and reproductive explanation is more likely than a social one.

A predilection for novelty and variety might be explained in some cases as a mechanism to avoid inbreeding. We know that as animals inbreed the number of abnormal genes in a population increases. Conception rates go down and the number of genetic anomalies goes up. It seems that most animals, including humans, have built-in mechanisms to move away from the family group, not to mate with close relatives—to marry out. In nonhuman primates, there's usually one sex that typically moves away from the natal group at sexual maturity. Most often, it's males who move, but sometimes females leave. Another way females might avoid the evils of inbreeding is to mate with those less familiar, peripheral males. It might also be a good idea to produce offspring with a various genetic heritage. If the environment changes, and several infants have all sorts of genes, at least one might make it. But it seems odd that females are drawn to these foreigners even when the resident males originally came from another troop. Perhaps they just aren't foreign enough.

Multiple matings might be best explained at its most basic level by the simple equation of sperm meets egg. Some have suggested that when multiple orgasm causes physical contractions of the uterus, the possibility of conception is somehow increased (Sherfey 1966). If this is true, then natural selection would push for females to mate with any number of males, and very often, just to help sperm transport. But there's no physiological evidence that orgasm improves sperm travel in primates—in fact the internal and external body contractions during female orgasm would more likely expel sperm than suck it in. A more reasonable physiological possibility is that female primates who gaily go about their mating business are buying sperm insurance (Milton 1985, Small 1988, Sommer 1989, Sommer, Srivastava, and Borries in press). Because ovulation has only about a twenty-four-hour window of opportunity, it behooves a female to keep her reproductive tract full of fresh sperm. The female obviously doesn't know exactly when the moment of ovulation occurs, so she must mate at every opportunity. This need is especially strong for females in groups in which many other females are in estrus at the same time. Under these conditions, with only a few males to service many females, sperm, not the availability of females, becomes the limiting resource. As an added bonus, any female who mates with a male depletes the sperm supplies for her troop sisters (Small 1988, Sommer 1989, Sommer, Srivastava, and Borries in press). Although sperm is theoretically unlimited, it's actually unlimited only in comparison to female eggs. Sperm is continuously produced by males internally, but it's not flowing from them like water out of a faucet. Sperm must be matured within the male reproductive system and stored until a reasonable amount is ready for ejaculation (Dewsbury 1982). When ejaculations are close together, the sperm count goes down accordingly (Michael and Zumpe 1978), and more time is needed in between bouts, a phenomenon called the latency to ejaculation. Primate males, both human and nonhuman, can't mate time and time again in rapid succession. They need time off. And this vacation might be frustrating to a female monkey in heat. Preliminary evidence suggests that this hypothesis may point to an important selective force in how female primates behave. In a study of Japanese macaques, only 34 percent of the mounting series actually resulted in ejaculation (Hanby, Robertson, and Phoenix 1971). And there seem to be fewer conceptions in groups in which there are fewer males (Dunbar and Sharman 1983, Silk, Samuels, and Rodman 1981).

Another sperm-limitation possibility refers back to a story about President Calvin Coolidge. Once on a vote-stomping tour, he stopped at a farmyard. While he was watching a rooster chase a hen, the farmer explained that when the male saw a new female his ardor was sparked by her novelty. Coolidge reportedly said, "Tell that to Mrs. Coolidge." Scientists have documented this phenomenon for male chickens, rats, and rhesus monkeys (Michael and Zumpe 1978). The evolutionary basis for the Coolidge Effect is simple: males should be excited when they see new females because those females represent new fertilizing opportunities. But no one has considered what the Coolidge Effect means to females, the *Ms.* Coolidge Effect. From the point of view of the female, she should always be the new girl on the block. If a male's sexual excitement, and thus potency, is highest when he experiences a new female, it's to the female's advantage to *be* that new female. Females want novel males as much as novel males want them.

Sexual Manipulation

All primates, including humans, use everything they have to stay alive. Our nonhuman primate relatives show us that under certain circumstances females use their sexual attractiveness to gain something. Biologists call this behavior "situation dependent receptivity." Nonhuman primate females, of course, have estrous cycles, and most species have an interest in sex only during proscribed periods. One might expect this cycle of sexuality to be under strict hormonal control and thus not accessible to the females themselves. Under special situations, however, female primates turn that sexual desirability on— and they most often do so to manipulate males.

Indian langur females, who sometimes live in small harems, use promiscuous behavior to save their lives and the lives of their infants (Blaffer Hrdy 1977). The langur harem is a shaky situation. A male holds onto a group of females only for a short time—twenty-seven months on average—and he's under constant pressure to keep out other males. On the periphery of his group lurks a cadre of males just waiting for the perfect moment to pounce and oust him. When a takeover occurs, the new resident male often kills young infants, thereby ending lactation for females and hopefully ushering in the next estrous period. Many nonestrous females faced with an infanticidal male immediately become sexual; they're facultatively promiscuous, switching on their sexual attractiveness. They bob and weave in a typical estrous langur manner, faking their receptivity, copulating their way into his favor. It doesn't matter whether they're pregnant: with or without an infant in tow, they swing into a sexual mode to bond quickly with the new male, and he doesn't kill their present infant.

A variation on the infanticide theme occurs among gelada baboons. These striking dark-brown animals live on the barren steppe of Ethiopia, where they scratch out a living on small grasses that grow on the dry land. Males dominate females and keep them in small groups. When one male takes over a harem group, the females show their acceptance of the male by presenting to him (Dunbar 1984, Mori and

Dunbar 1985). Males don't kill infants, but females often automatically abort fetuses and then mate with the new males; it's as if they were cutting bait and starting again.

Bonobo females also manipulate males, but in a more peaceful manner, and their gains are usually something much more tangible—food. Frans de Waal filmed a scene at the San Diego Zoo in which a male bonobo hoards a large clump of branches. He moves up to a female, presents his erect penis by spreading his legs and arching his back. She rolls onto her back and they copulate. In the midst of ecstasy, she reaches out and grabs the leafy branch from the male. As he pulls back, finished and satisfied, she moves away, clutching the branch to her chest. But this is no simple matter of the female's trading sex for food. She can get away with the trade because he and she are of equal status. Bonobo researchers believe that the extended swellings of female bonobos and the fact that they have sex any time and anywhere is what makes the relationship between males and females equivalent. To the unknowing eye, the scene described above looks like manipulation, but it's really just an agreement between two individuals on what they want out of a situation.

The female estrous cycle can be a hotbed of manipulation, but certainly there's nothing conspiratorial about female physiology. Although estrous signals sexual receptivity, it doesn't always signal fertility. For example, many nonhuman primate females exhibit estrous cycles when they're already pregnant (Blaffer Hrdy and Whitten 1987). This condition is very common among macaques. In my group of Barbarys, 79 percent of the females cycled again, with huge swellings, after they had conceived (Small 1990b). To the practiced eye, these postconception swellings are a bit smaller, but one would have had to get out a tape measure to confirm that subjective opinion. The infertile cycles are also somewhat shorter in length. But to males, these pregnant females are just as good as ovulating females. Barbary males approach females just as often as they do when females are having fertile cycles. And although female interest wanes some and they don't present to males as much, males copulate with them just as frequently, or sometimes more often, even though they're already pregnant. Primatologists aren't really sure why females waste energy on growing a reproductively useless swelling, or why they bother to copulate with males when a sperm deposit is unnecessary. Some suggest that these "extra" cycles are merely a result of a hormonal system that has trouble shutting down (Bielert et al. 1976). Others suggest that females have a second cycle to confuse males about paternity (Blaffer Hrdy 1977), but if this were true, there might be more than one postconception estrus, and the pattern would be more common. And a third group suggests that females are manipulating their fellow females by copulating and depleting sperm supplies for others (Small 1988, Sommer 1989, Sommer, Srivastava, and Borries in press). No matter the ultimate reason for their appearance, these pseudocycles do at least seem to be an option open to females.

The Way of Primate Females

While none of these explanations is straightforward or necessarily clear, each does suggest evolutionary reasons why female nonhuman primates are so interested in copulating with different males. We may never find a universal explanation for the primate urge to seek out new males, but the studies I've described here suggest some interesting possibilities. More important, the behavior and the theoretical explanations about the sexual assertiveness of female nonhuman primates pulls these females out of the negative light of promiscuous sex-crazed harlots into the arena of strategizing creatures trying to improve their individual reproductive success.

And finally, the main concern of primate females is conceiving. Contrary to past notions that females will be selective in this endeavor, predisposed toward only the finest males available, there's evidence that females are more concerned with becoming pregnant in the first place. It's not that the contribution of male genes is unimportant to females but that females may be more concerned with ensuring conception than with conceiving with a particular male. The constraint of a few hours or days around ovulation dictates a simple solution: mate with many males, many times, don't delay.

Primatologists have empowered primate females by acknowledging their sexual assertiveness, but we often stop short in accepting the fact that sexual assertiveness may result in less than choosy behavior. As a passionate female with reproductive interests in mind, a female must do what a female must do. This assertiveness is especially important for females in multimale groups that breed seasonally and that might not have enough sperm to go around, or when a male of a harem group is not up to the task, or when the bonds of monogamy leave a female with only one male. After satisfying the prime motive of conception, females may then have the secondary luxury of selecting preferred partners.

More than any other variables that stand out to our human eyes, novelty and variety, appear to be the preferences of female nonhuman primates, and as a consequence these females are perceived as promiscuous. If nonhuman primate females ever had a sexual revolution, it was for the right to exhibit less-than-choosy behavior. Lowering their standards in the interest of conception, and taking anyone who is interested, must then be one of the innate abilities of primate females.

A simple question then follows—how close is this pattern to human female behavior?

CHAPTER 7

Human Female Sexuality and Mate Choice

THE LINE OF DECORATED FACES undulates slowly toward the crowd. Each willowy figure stands on tiptoe, straining to appear the tallest member of the group. Teeth flash in a frightful grimace, and eyes roll back to expose as much white as possible. The streak of yellow dye down the nose, the purple-powdered lips, and the jangle of multiple strands of beads hung from headdresses, all add to the surreal atmosphere. These are Wodaabe tribesmen of the Niger, and in a twist of the usual sex roles, the men of this culture often gain wives by wooing them with decoration and dance. This dance, called the *yaake*, is part of a festival or *worso*, in which unmarried Wodaabe women choose mates (Beckwith 1983). And they do so based solely on male beauty.

The teen-age !Kung San girl isn't ready for marriage, according to her. She's displeased by her parent's choice of a man fifteen years her senior, someone her family wishes to form an alliance with. The day before the wedding, her mother and two older sisters carry her to the hut specially built for the couple (Lee 1984). She screams and kicks, letting the entire Botswana village know of her anger, and at night, when she's supposed to lie between her future husband and her sister, she sneaks out and sleeps in the bush. Her protest is finally recognized as real, and her parents call off the wedding, knowing that if they force her the marriage is likely to end in divorce anyway. A year later, her parents choose another man, and again she's carried to his hut. This time she also screams, but the protest is a fake, because she approves of this husband. But she kicks and yells anyway, a ritual acknowledgment of the tension among !Kung San men and women and their families at the time of marriage.

An American man is introduced to a woman by a close friend. He asks her out for coffee and they discover that they have a lot in common. He thinks that she has a beautiful face and she thinks that he's funny. They make a date for a movie later in the week, and that too turns out well. Within a week, the couple have a sexual relationship, which cements their attachment even further. Later that year they move in together, and some months after that they decide to marry. The choice is based on love, lust, and companionship.

The details of human mate choice are as variable as the humans who make the choices. We actually mate in two general ways, sometimes very separate ways. In marriage, mating occurs within a legal, social, and economic bond. This bond is acknowledged by the larger community, and it has certain expectations. The other kind of mating, outside of marriage, also occurs regularly in all human cultures. In both arenas, genes can be passed on when children are conceived. From an evolutionary standpoint, then, we might expect mating and marriage choices to be subject to the rules of natural selection. Do our mate preferences make evolutionary sense? Do we use different criteria for our temporary sexual partners than we do for potential spouses? Are women passive in the decision for mates and spouses, or are they strategizing creatures like their nonhuman primate sisters?

It's much easier to look at marriage choices than lovers. Marriage, after all, is a long-term arrangement that carries significant consequences and it is usually documented by a society. Most cultural anthropologists suggest that culture holds all the cards in the arena of marriage. The evolutionary urge to pass on genes to the next generation, they maintain, is obliterated for humans by the heavy blanket of culture. Obviously, marriage in every culture is an economic and social phenomenon, not just a way for genes to be proliferated, not just mating. The myriad types of marriages around the world support the notion that cultures have their own agendas where marriage is concerned. But the issue is not that simple. Even when human marriage is an economic or political arrangement, the production of children is of prime concern to all parties involved. Although there are strong differences between human marriage and nonhuman primate mating, there's also the common fundamental issue of producing and raising offspring. In other words, the basic function of human marriage is to form a unit in which children can be produced, and all cultures, families, and individuals are concerned with producing children. Underlying the heavy hand of culture, therefore, humans share with their nonhuman primate cousins the primate urge to make babies. Humans just have children most often within the bonds of matrimony, a particular kind of mating system.

Human mate decisions outside of marriage for reasons other than procreation are more difficult to evaluate. Humans have disconnected sex from reproduction, and human females are often receptive to sexual activity during all times of the cycle. Birth control in our modern age reinforces this disconnection even more by removing the fear of pregnancy. In fact, human females, like some nonhuman primates, such as bonobos, use copulations, in one form or another, for nonsexual reasons. Sexual activity can occur among friends, can be used to form alliances, and can change the power of female roles. I've established that nonhuman primate females are assertive sexual beings, mating with many males and trying as best they can to exert choice. Because we share with nonhuman primate females many of our broad patterns of behavior, we may also share our sexual nature. Presumably this sexuality influences and interacts with the development of human marriage and mating practices.

This chapter asks, is the layer of culture so thick that we humans approach sexual activity, make mate choices, and accomplish reproduction differently from other animals? More specifically, I ask, how have human female sexuality and reproductive interests influenced the evolution of the human species?

Biological Ties That Bind

Mating and reproduction are the keys to understanding individual reproductive success; this is as true for humans as it is for monkeys. Although culture has presumably had a great impact on the expression of human sexuality, it's also reasonable to suggest that human sexuality, mating, and marriage practices have been molded, in some degree, by evolution. Humans don't mate at random, and their choices of mating and marriage partners do result in differential reproductive success regardless of what culture dictates. If we're to find any biological roots in human behavior at all, they should most readily appear at this basic level, during mating, sexual expression, and marriage, when genes are passed on.

Human female sexual behavior and mate choice, the focus of this chapter, is extremely difficult to approach. Women, unlike animals, may be able to tell us how they feel and what they do, but those same human-specific abilities give us only part of the picture, and only a handful of studies have focused on female sexual response. Further, attitudes about female sexuality are deeply embedded in cultural morality, and this is true for our own as culture as well as for cultures foreign to us. Most women won't easily discuss their sexual history or their attitudes about sex, and every women is influenced by what she reads, sees, and hears in the media to give the "right" answers (Faludi 1991). Cross-cultural reports of human female sexual and marital patterns are also suspect because sexuality is sometimes impossible to study by an outsider. And then, as any social researcher knows, what women do and what they say they do are often two different things. The influence of female choice on the evolution of our mating practices is even more difficult to discern, for there's no behavioral record. It's hard to know why we mate a certain way today if we don't know the road that brought us here. The ancestral nature of human female sexuality and of mating in particular is really only scenario building based on speculation and pure fantasy.

But there's no question that human female sexuality, marriage, and mating all display certain very broad universal patterns. Physically, women have few external signals of ovulation and they are able to participate in sex during all phases of the cycle. Women also have a predictable pattern of sexual response with orgasm. Emotionally, women in all cultures form bonds with men, marry, and have children. Individual female concerns over children and family are not necessarily opposed to the concens of those making the rules of family construction. And most important, women participate in their own destiny, even if their participation is simple acquiescence. These universals underscore a sisterhood that has been molded over millions of years during the evolution of our species.

The evolution of our particular human female biology has also caused a major problem for males: there's no way for a man to know for sure that he is the father of a particular infant because the time of ovulation is obscure. And yet mothers of highly dependent nursing babies need help in terms of food and infant care. In addition, only human children continue to be nutritionally dependent as juveniles; they no longer nurse but they don't exactly feed themselves either (Lancaster 1989, 1991). Fathers who stay attached to a mother will be greatly in demand in a system in which women usually have several nutritionally dependent offspring. But a male sees this problem differently—he will care for an infant only when he knows the needy child is his own. As a consequence, societies, families, and men often go to great lengths to ensure paternity. Women need men to aid in rearing children, but men must be assured that his aid is going into the correct packet of genes. The female need for aid to improve her reproductive success and the male response that he will give only when paternity is clear have dominated the evolution of our species and have colored the myriad ways men and women have compromised to rear the next generation.

Did Lucy Choose Mr. Lucy?

We have no good fossils from the period when the human lineage diverged from that of the apes. From 8 to 3 million years ago, the human record is a sparse group of bone fragments that tell us next to nothing about how our ancestors looked. The first clear human, or hominid, appears around 3.5 million years ago in the form of a species called *Australopithecus afarensis.* Although most scenarios of human evolution refer, speculatively, to the gap before *afarensis* as a "protohominid" period, the first really hard clues to human evolution begin with *afarensis.* Here we see a small creature. The most famous *afarensis* is Lucy, 40 percent of a full skeleton, found by Donald Johanson and colleagues in 1973 in Hadar, Ethiopia (Johanson and Edey 1981). Lucy has a special charm. Her name, derived from the song "Lucy in the Sky with Diamonds," emphasizes the importance of her fossilized bones: they're true gems. She was a small member of her species, only three and a half feet tall, but she has taught the paleontological community much about her group. Her tiny pelvis, like one disconnected from a large doll, demonstrates that she was fully bipedal. In some ways, Lucy and her colleagues were the true missing links between apes and humans. She and the other *afarensis* had brains the size of chimpanzee brains—they were none too bright but walked like modern humans. Some had pointed premolar teeth like apes' and larger-than-human canines. And their curved hand and foot bones indicate that they may have spent some time in the trees.

What we know about Lucy's love life, or that of those who went before her, is nothing, or next to nothing. Because infant brains were small, we can assume that childbirth may not have been as difficult for Lucy as it is for modern humans. It's also reasonable to assume that Lucy and her female compatriots didn't have genital swellings—bipedalism makes that unlikely. We don't know, however, whether they had other signs of ovulation, or had pendulous breasts like modern women's, or how often or with whom they mated. The mating system is almost a complete blank. The only indication of what might have been going on between the sexes comes from the difference in body size between males and females. Male *afarensis* are much larger than females. One thing we know with confidence from other primates is that differences in body size between the sexes indicates a polygynous mating system (Clutton-Brock and Harvey 1976). Males evolve larger because big bodies help them in male-male competition. Male and female members of monogamous primate systems, in which there is little male-male competition, tend to be about the same size. Thus Lucy was probably not living in a monogamous social system. She may have been in a harem situation, but it's equally possible that she had access to several males and that they had access to her.

Lucy heralded in the era of Australopithecines, of which there were at least four different species roaming the woods and savannas of Africa. We're not sure which *Australopithecus* was responsible for carrying on the human lineage, but we know that about 1.5 million years ago a distinctly different creature arose, one that shares our genus *Homo*. This group had extremely large brains and they made tools. It would be a mere million years before this genus and it's decendents began to move out of Africa and populate the earth. *Homo habilis*, *Homo erectus*, *Homo sapiens neanderthalensis*, and *Homo sapiens sapiens* are the species and subspecies that make up the human lineage down which human females evolved. The fossil evidence tells us that sexual dimorphism, the difference between males and females so apparent for *afarensis*, gradually became less and less. Archaeological evidence shows that small human groups lived together, worked together, and played together. But what none of the fossils and none of the artifacts tell us is how they loved or how they mated. Our mating and sexual history is nonexistent and we must rely on a mix and match of clues.

In that sense, scenario builders who weave complex tales of our evo-lutionary past are like mystery writers who know the end of the story, have a few hints about how the event occurred, but no real evidence. As a result, the stories of human mating evolution are as varied as the storytellers.

Stories of Female Evolution

In the story of human evolution, as told by anthropologists, the female role has often gotten short shrift (Fedigan 1986). In the 1960s the most popular scenario outlining how the human social and mating system evolved focused on males as the central figures in human evolution. And this human male prototype was a burly hunter who went out on dangerous expeditions, used sophisticated tools, and cooperated with other men to gather meat for a family waiting back at camp. "Man the Hunter" is, perhaps, the most popular, and still the best-known, scenario of how the first humans lived (Lee and DeVore 1968); this is the caveman hunter that most nonscientists see as our ancestor. It's also derived from a male-biased view of human evolution. According to this scenario, women apparently didn't even figure into the development of the species. As Linda Fedigan, a female anthropologist, points out, these early scenarios suggested that only man the hunter was responsible for the invention of everything social and technological about humans (1986). Women were just dependent persons left at home waiting for a good meat meal that only men could provide, and this meat was an essential ingredient for the only role women occupied—baby makers. Everything that defines hu-manness, from bipedalism to high intelligence, was ascribed to men, and women received human traits only on what Fedigan calls men's "coattails"; women are bipedal because men need to be, women are intelligent because men need to be, etc.

In a kind of backlash reaction, several female anthropologists soon pointed out that women in today's hunting and gathering societies gather the foods that are the stable core of their diets, and this practice may reflect the type of subsistence utilized by our ancestors (Dahlberg 1981, Linton 1971, Tanner 1981, Zihlman and Tanner 1978). Given this fact, "Woman the Gatherer" must be figured into any scenario of hominid evolution. There's no reason to assume, for example, that the tools used for hunting are any more complex than the tools needed for gathering or that hunting requires more social cooperation or intelligence than food gathering. The most important result of this backlash was that most anthropologists realized that females make up half the species and that the female part had been left out of the story, that female strategies and interests are certainly as important to the evolution of the species as are male strategies and interests.

Even with the enlightenment provided by the female anthropologists who promoted "woman the gatherer" in the late 1960s and early 1970s, the male-centered model continues to be incredibly tenacious. When Sarah Blaffer Hrdy published her book on female primates in 1981, almost twenty years after the first man-the-hunter model, she called it *The Women That Never Evolved*, and this title turned out to be prophetic. That same year, the prestigious journal *Science* published a complex model of human evolution by the hominid paleontologist Owen Lovejoy in which males serve as the central figures in human evolution. Women once again stay at home and serve males with sex while males hunt for food. The paper is appropriately titled "The Origin of Man"; women just don't evolve, apparently. Even in our politically correct 1990s, when authors try desperately to see both the male and female roles in evolution, biases still slip in. In a recent book on human nature, *The Third Chimpanzee*, published in 1992, Jared Diamond explores the evolutionary reason for the large size of the human male penis. He postulates that large penis size could have evolved for functional reasons—to make copulation in various positions possible. But since apes, namely orangutans and bonobos, have sex in all sorts of ways with shorter penes, Diamond correctly says that this isn't a likely explanation. He then suggests that a large human penis would do well as a signal, but that the signal certainly isn't directed at females—modern women don't like to look at photographs of penes, he claims. The large human penis, he then concludes, evolved as a display among men, as result of male-male competition. What Diamond fails even to consider is the possibility that the large human penis might have evolved for female pleasure. Back in 1979, the renown physiologist R.V. Short had dismissed male display as a possible explanation for penis size, because copulation among humans occurs in private and males in all societies cover their penes, sometimes with constricting sheaths (Short 1979). Following this lead, anthropologist Helen Fisher wrote that the *only* function of male penis size could be female pleasure; the large penis pulls on the outer third of the vagina and facilitates clitoral stimulation (1982). Remarkably, Diamond, who cites Short's paper, simply ignores this idea. He doesn't even consider the possibility that female sexual pleasure might be an important selective force, even of a male trait that's exaggerated and could fit the criteria for selection by female choice. Although neither Diamond, Short, nor Fisher can be proved right, it seems reasonable to think through what advantages or costs occur for each sex in these scenarios. Ignoring either sex dismisses half the species and leaves evolution to work in a lopsided way.

Several current models of human evolution take into account both male and female strategies. But instead of hunting and gathering as the force behind human social patterns, sex and parenting are in the spotlight. Several authors suggest that early in human evolution female humans "used" their sexuality to gain male parental care or male resources and goods that help bring infants to maturity (Alexander 1990, Alexander and Noonan 1979, Benshoff and Thornhill 1979, Irons 1983, Lovejoy 1981, Strassmann 1981, Turke 1984). No longer bystanders in the story of hominid evolution, females are the perpetrators of the basic human social system. And this system, they all maintain, is monogamous, or pair bonded. Their evidence for monogamy as the ancestral and present-day "natural" condition for humans is rather questionable. In the first place, there is extensive variation in the ways in which humans mate and marry. Jane Lancaster, in particular, views human mating systems through an ecological looking-glass and points out that mating flexibility, not rigid monogamy, is a mark of our species (1989, 1992): polgygny appears in the vast majority of cultures, there's a sprinkling of polyandry around, and single-female households are a growing pattern of family formation. In general, as humans moved from a hunting and gathering way of life to more sedentary communities, family formation patterns have responded in flexible ways. Lancaster maintains that human female reproduction, the probable motivator of family formation, is unusually sensitive to the resources needed to produce and rear children and that even single-mother households so universally prevalent today are just another of women's response to changing ecological conditions (1989). Tenuous bonds between men and women will appear, she contends, when paternal investment is low. A woman might not opt for, or be able to rely on, a tight bond with a single male who might invest enough in her children over her lifetime, and she will therefore seek community or family-member contributions to provide for growing children. Women are at their most vulnerable when they must depend solely on men for the resources necessary for producing children and getting them through the juvenile period (1991). This vulnerability is pushed further because a woman most often has more than one child at home, needing her attention and dependent on her for food; children and juveniles rely on adults for almost every nutritional need, even if it's not breast milk. But women can also sometimes manipulate their own reproductive biology to fit changes in parenting resources by increasing or decreasing the interval between births, killing infants they can't readily care for, or by investing only in those with a bright future (Lancaster and Lancaster 1987). Given this perspective, there is no natural, original, ancestral mating or parenting system for humans, but a flexible system within which both men and women operate to bring up the greatest possible number of offspring.

But those who believe that monogamy is the ancient human condition are committed to the superiority of one-on-one bonds that keep families together for the good of passing on genes to the next generation. Infants are so dependent, they suggest, that by definition two parents are required. But the requirements of infant care don't seem to be enough to keep parents bonded. Instead, these authors see a constant tension between females, who need care for their offspring, and males, who want to run off and copulate with other females. Lovejoy goes so far as to suggest that the pair-bond evolved from a situation in which the protohominid female stayed close to a home base to deal with the dependent infant while the mobile male wandered off in search of food—a scenario that sounds strikingly like the picture of a stationary egg waiting around for a mobile sperm (1981). Female sexuality, they all suppose, is the bait that females use to keep males in tow, and thus sex is the basis of the tenuous bond between men and women.

In all these scenarios, concealed ovulation and so-called continuous receptivity are the keys to understanding the mating relationship and the eventual social system of hominids. Authors hypothesize that somewhere along the human lineage external displays of estrus were selected against in females. Most scenario builders maintain that this concealment is from females and males alike (Alexander and Noonan 1979). This is an important point because masked ovulation leads to all sorts of sexual confusion. If a female doesn't know when she ovulates, she must mate with males all the time if she wants to conceive. And if a male doesn't know when a female ovulates, he's an easy mark for cuckoldry. But no one has really sufficiently answered the primary question in the concealed ovulation story—is it really concealed, and if so, from whom? Nancy Burley, a biologist, has suggested that concealment is really aimed at females—if females knew when they ovulated, she supposes, they'd never conceive and bear infants because of the high personal cost of raising children (1979). So-called concealed ovulation is then evolution's little trick to make females conceive by disguising the time conception is most likely. Some women, however, claim to know when they ovulate. In a pilot study of two hundred college women, I found that 67 percent said they thought they knew when they ovulated (Small ms.). (When I asked males who had been in long-term relationships whether they "knew" when their partners ovulated, those few who answered "yes" said that they knew because their girlfriends told them.) In other words, concealed ovulation in humans is certainly not a settled fact. And although how we view ovulation today may have only superficial bearing on how our primitive ancestors experienced it, understanding the consciousness of women and men today about this fundamental event of reproduction would be a useful piece of data to add to hominid scenario building. In the meantime, there's no reason to assume that hominid females necessarily "lost" anything.

Regardless of the place of concealed or revealed ovulation in human evolution, other labels have been affixed to modern humans which many researchers suggest indicate a particular path to our history. For example, as I have pointed out, human females are often considered sexually "continuously receptive," meaning that females are always ready for sex and that this receptivity is supposedly a major feature of human female sexuality both today and ancestrally (Lovejoy 1981). Some authors have acknowledged the difference between humans and almost all other primates in this feature by regarding human females as experiencing a kind of low-key continuous receptivity (Alexander and Noonan 1979). The term "receptivity" (correctly or incorrectly) has been taken from Beach's definition of estrus for mammals (1976). As I mentioned in Chapter 5, calling human females "continuously receptive" is inappropriate for two reasons. First, it may be inaccurate to use the term "receptive," which defines an estrous cycle, for an animal (a human) that experiences menstrual cycles. Second, human females may be continuously *attractive* (that is, males are usually interested), but they're certainly not continuously proceptive (meaning they initiate sex) or receptive (always willing to help in the process of copulation with anyone). Human females have been portrayed alternately as acquiescent sexual beings who submit to male attentions or as manipulative sexual partners who have gained male care by trading sex for favors (Blaffer Hrdy 1981a). Underlying both these different views is the suggestion that human females are continuously ready for sex and that they can use that readiness when they please. Although there may not be clear answers to questions about the "true nature" of female sexuality, it's perhaps reasonable to suggest that females are neither passive receptive vehicles nor continuously ready to have sex (Adams, Gould, and Burt 1978, Gebhard and Johnson 1979); both these descriptions are extreme. The most we can say is that human females today are almost always attractive to males, although not all females all of the time, and they also initiate sex sometimes, and are sometimes ready for sex. In fact, Beach's three categories of sexual behavior describing female sexuality of other animals—attractive, proceptive, and receptive—don't really fit human females anyway, and consequently shouldn't be used to describe the evolution of human female sexuality. The best we can say is that human females are physically able to have sex during all points of their cycle because their sexual motivation is less tied to changes in hormones than it is for females who experience estrus. Just because human females don't have episodic periods of heat, however, doesn't mean they're continuously hot—nor does it mean they're continuously cool.

In any case, these authors suggest that, along with acquiring concealed ovulation, our hominid ancestor females became continuously available for sex and that they used their sexual flexibility in trade for faithful mates, paternal care, and reciprocal goods and services. Females were able to make this trade because they had something males wanted—sex—and were always able to give it. The biological target of this trade is the inexhaustible male sex drive that is coupled with improved reproductive success for males who mate with multiple partners and disseminate sperm at every opportunity (Symons 1979). Females then could trade their sexual access to males who would do anything to copulate. There are, however, several major flaws in this line of reasoning. First and foremost, it suggests that for generations hominid males, who would supposedly have been most reproductively successful by mating with multiple females, had been fooled into monogamy when a female offered herself as one continuous sexual outlet. This male turned from a strategy of lots of sex with many females to lots of sex with only one. But this shift doesn't make evolutionary sense. Monogamy based on male sexual satisfaction with one ever-available female couldn't easily evolve because staying with one female and filling her with gametes all day long gains him less reproductive success than if he were to roam a bit. There's also no clear evidence that human males naturally have higher sex drives than do human females or that males are more interested in extrapair matings than are females. Data on current humans (see later) indicates that females are just as sexual as males and that if society stopped its interference females would enjoy themselves sexually as well as males. These criticisms undermine the notion that human males are compelled toward a life of wild sexuality and that they can be restrained only by women.

The reasoning also falls short from the female perspective. The human female really isn't necessarily continuously receptive and available for sex, so how could she have lured a male? And what happened to female orgasm in all these scenarios? They suppose that only males enjoy sex and that hominid females use it only to keep a male home. But here's a creature capable of multiple orgasms who will repeatedly search for sexual outlets. Female orgasm perhaps puts the pair-bond in jeopardy. The female might find sexual satisfaction with one male, and mutual pleasure could cement the male-female bond, but it's just as likely that she won't be satisfied with one male (as he might not be with her) and seek sexual pleasure elsewhere. Lancaster, seeing sex as one of the most powerful biological stimulators, maintains that human females play as significant a role as males do in forming human pair-bonds (1979). She believes that female sexuality, and the disconnection between sexual motivation and the hormones of reproduction, empowers females rather than places them in a position vulnerable to male domination. The pair-bond is maintained by both partners, not by one partner's fooling the other.

More important, the sexual drive may not be at issue at all for humans. Compared with other primates, humans are not especially sexual creatures (Short 1976). Unlike nonhuman primate males in multimale groups, human males have small testes and low sperm counts. Also, Ford and Beach maintain that human females have long periods of adolescent sterility, and they are often pregnant or lactating and abstain from sexual activity. Most societies, they say, also have cultural taboos against sex during much of the year or over parts of an individual's life span (1951). We like to see ourselves as highly sexual creatures, and if this were true, perhaps human females could wield their sexuality as a weapon. But the effect of the weapon is only as good as the target, and thus sex as a weapon may actually have little power in the daily machinations of human societies. And finally, there's no necessary causal link between concealed ovulation and monogamy, or monogamy and frequent sex. Other primates with concealed ovulation, such as vervets, aren't monogamous in any sense, and many monogamous primates, such as gibbons, rarely copulate. There must be other evolutionarily compelling reasons for males to remain with one female, and females to stay with one male.

Those who argue that females need to snare males into paternal care by giving them sex overlook the fact that the human males' reproductive success is as dependent on parental care as is female success. If a male abandons his offspring, it's likely to die. Thus females need not coax males to stay and help, selection will develop this behavior anyway. There are actually two levels that humans must deal with to pass on genes, and these two levels are in conflict. At the level of mating, a male should try to inseminate many females, but in terms of parental investment, he needs to invest in offspring. For a female too, there's a conflict between mating and parenting. She might want to mate with several males to ensure conception or to try a nonbonded male for his different genes. But as a parent, she too must devote her energy to the dependent human infant. The conflict between human males and females is emphasized at the level of mating, but compromised at the level of parenting. Conflict also arises for any individual between levels of sex and parenting. The individual who compromises those levels does best in the game of reproductive success.

Perhaps the most reasonable scenario for the evolution of human mate choice and family structure is one that has little to do with sexuality. After all, humans have virtually disconnected sex from reproduction, and it would be hard to imagine that, given this disassociation, sex would figure so high in the social system of ancestral (or present day) humans. William Irons and Jane Lancaster have independently suggested that the human social system, the pair-bond, probably evolved as one of reciprocity (Irons 1983, Lancaster and Lancaster 1983). This exchange system was probably not built on sexual desire, as other authors have suggested. Both parties, males and females, cooperated long ago to make and raise dependent offspring. Just as pair-bonded gibbons or titi monkeys, who sometimes copulate outside the bond but return to maintain a patch of forest with their established partner, humans separate mating from the daily needs of maintaining a family unit. Protohuman males might have hunted and contributed meat while females gathered and contributed vegetable matter to a common family pool. Or there might not have been a significant division of labor in the human lineage until rather recently, during the *Homo erectus* stage, and thus the early humans may have worked together in similar ways on similar resources to support the family or group (Fedigan 1986, Leibowitz 1983). Beyond the goods necessary to keep infants and parents alive, both parties also gained something else in a reciprocal manner—sexual satisfaction. Pair-bonded males must be assured of paternity before they invest, and presumably most mating occurred in the context of the unit. But because most human sexual activity was, and still is, nonreproductive, quick matings outside the pair may have been inconsequential to the bond within which children were nurtured. This isn't to suggest that the compromise bond is a nice easy solution to differences in reproductive strategies. If anything, such a bond causes continual friction within a relationship, and over time for any individual. R.V. Short calls this a compromise between "our polygynous nature and the need to maintain a parental pair-bond for the benefit of children" (1979). In support, the data on extramarital affairs in Western and other cultures suggest that humans today usually see marriage and sex, spouses and lovers, sex and reproduction as different entities.

The result is a pair-bonded social system, with most sexual activity occurring within this bond, but not exclusive of mating outside the bond. Thus the origins of the human mating system may not be strictly rooted in monogamy, although the origins of our social system are strongly pair-bonded.

Sex and the Female Hominid

There are few indisputable facts about the sexual and mating behavior of *Homo sapiens* females today. The only thing we know for sure is that human females have no sexual swellings or extreme signs of ovulation and that sexual interest is not as episodic as it is in estrous females. There's some evidence that the presence of a male (not copulation per se) improves the incidence of ovulatory cycles in modern women (Veith et al. 1983): college women who spend several nights sharing a bed with a male, regardless of sexual interaction, ovulate more regularly than other women. There's also speculation that human females who live in close quarters cycle synchronously (Adams, Gould, and Burt 1978, McClintock 1971, Russell, Switz, and Thompson 1980). Some say that human females copulate at all phases of the hormonal cycle, while other debate the subtlety of female sexual interest relative to cycle phase and during pregnancy (Adams, Gould, and Burt 1978, Gebhard and Johnson 1979, Kenny 1973, Udry, Morris, and Walker 1973). Females may, in fact, have their strongest desire for sex, and even signal this desire to males, during midcycle or right before menstruation (Adams, Gould and Burt 1978). But no one really knows how present-day females conduct their sexual lives relative to ovulation. Surprisingly, there's little information about how, when, and why human females participate in sexual intercourse. And yet, as I have discussed, there seems to be a prevailing culturally bound notion in Western writing, both popular and scientific, that human females are less interested in sex, or sex with multiple partners, than are males. For example, anthropologist Donald Symons has suggested that strong differences exist in human male and female sexual practices and that these differences are biologically based (1979). He proposes that a desire for sexual variety in partners in human males is an evolutionarily selected strategy as it is for other males. He uses literary references from Western culture and cross-cultural data to support his hypothesis. When a female does avail herself of different partners, Symons believes, she does so to improve her chances of conceiving with a new male, one better than her husband. Males, on the other hand, desire variety for variety's sake alone. The difference in the investment that males and females give to infants results in different mating tactics even today. Human males are carefree about sex, he maintains, interested in fooling around with any number of females, and females are choosy because of their high parental investment and are not particularly interested in sex.

Symons words are echoed by Martin Daly and Margo Wilson, two human sociobiologists who have written extensively about the biological basis of human behavior (1978). They refer to expected differences in human male and female sexual strategies as "the reluctant female and the ardent male" or "the nurturing female and the prodigal male," and they suggest that sexual roles are set by evolution. There are, however, major flaws in Symons's and Daly and Wilson's application of differences in parental investment to the human case. First, as I mentioned before, the dependency of human offspring has reproductive consequences for both partners, not just females. Although some human males may benefit from copulations with many females, most males benefit by contributing to offspring for which they have confidence of paternity (Frayser 1985). And if a female has a say in which male she might copulate with, the same evolutionary framework that suggests human females will be choosy also suggests that such choosiness will select against males who don't demonstrate commitment to offspring. Second, human males actually have extremely low sperm counts and slow sperm mobility compared with other primates (Small 1988). Any male's potential for conceiving with many females, if they don't have external signs of estrus, is relatively low. Third, Symons's current data on human polygyny—in which a male marries many females—really doesn't support the notion of a biologically wired urge for sexual variety (Frayser 1985). Although he believes that a tendency toward polygyny in some cultures supports his view that men are hard-wired to want many sexual partners, the function of polygyny worldwide has little to do with sex. Polygyny occurs significantly most often in patrilineal, male-based societies in which male-male alliances are an imperative; males marry a number of females to form alliances with the females' kin. In other words, plural marriages, in which males are still restricted to a few females, are the result of social or cultural factors, not sexual desire (Frayser 1985). The polygynist might have more than one mate, but he's still stuck with the same two or three wives. And fourth, Symons's use of a Western model of male freedom and female restriction might make intuitive sense only because it comes from our own culture, in which males are bombarded with pictures of women of every shape and size used to sell consumer goods. No one has yet really tried the same tactic with women—we don't even know if the sale of female goods would improve if nice-looking males were attached to the products.

Data from Darwinians who are more female-oriented contradict the notion of women as reluctant, choosy sexual partners and provide an alternative picture of human females as naturally sexual creatures but restricted by social norms. As the previous chapters have shown, nonhuman primate females are not passive or choosy. They are sexually active and interested in a variety of males when males don't restrict their behavior. The sociobiological theory that females should be choosy and careful isn't clearly born out by the data, at least for nonhuman primates. If our nonhuman primate cousins are so sexual and sexually assertive, can human females be so different?

It's estimated that in over three-quarters of the world's cultures the male and female sex drives are considered equally strong (Whyte 1978). But generally, societies restrict female sexuality more than male sexuality. The constricted female behavior does not mean that these women are biologically less sexual than men, only that a repressive society doesn't allow them to express their sexual desires (Frayser 1985). In fact, most cultures restrict females *because* they see female sexuality as something that must be restrained (Blaffer Hrdy 1981a, Broude 1980). The missing piece of the female sexuality puzzle is females themselves—what do women say about their sexual desires and motivation, and how does that affect their mating and marriage practices?

If the data on nonhuman primate sexuality is rather scarce, the information on human females is virtually absent. There are, however, some studies of human sexual practice which are informative, but they focus only on American culture and thus can't be considered indicative of the full nature of human female sexuality. The first extensive study was conducted by Kinsey and colleagues from 1938 until 1950 (Gebhard and Johnson 1979, Kinsey et al. 1953). Their work is based on interviews with 5,940 women of all ages, socioeconomic classes, education and religious backgrounds, but only of Caucasian ethnicity. Kinsey was interested in the physiological and psychological aspects of human female sexual response and the changes of those responses with age. This study is quite remarkable for its time; long before the so-called sexual revolution, the female subjects answered questions about orgasms, coital frequency, pre- and extramarital coitus, and masturbation. The findings were, and still are, important in breaking cultural stereotypes about female sexual response. For example, using data on masturbation, they discovered that human females reach orgasm as quickly as males and that only the position of the clitoris relative to the vaginal opening makes penile-induced orgasm sometimes difficult for females. They discovered that half of the females questioned had engaged in intercourse before marriage. Of those, 41 percent had coitus with her fiancée and a least one other male before him. During marriage, female interest in sex rises and remains steady while male interest declines with age. Although Kinsey believed that the rate of sex with different partners by female informants was lower than that for males, females also often engage in sex with more than one partner. The incidence of extramarital affairs confirms this fact. Gebhard and Johnson, who reevaluated and added to the Kinsey data, report that 23–56 percent of their males and 17–25 percent of their females had been involved in extramarital affairs (1979). And more interesting, the possibility of questions about extramarital behavior was the single most important reason potential subjects decided not to take part in the study. It was also the question that received the highest "refusal to answer" rate (Buss and Schmitt in press).

The second major sex survey was sponsored by the Playboy Foundation in the early 1970s (Hunt 1974). This study was based partly on 2,026 responses to a lengthy questionnaire sent out to a random sample of Americans. Also, people in twenty-four cities across the country were telephoned and asked to participate in small private discussions about sexuality, and one out of every five persons agreed to meet with the survey team. They too then filled out the questionnaire, so that this sample was more random in nature than those people who chose to answer a mailed form. The author and his wife also interviewed individuals to gain a sense of the meaning of the results of their data. This study seems to concentrate mostly on married individuals and says little about single women, except to call them all "premarital." But it does track trends since the Kinsey study. It found that the percentage of women having sex in their teens had rapidly increased since Kinsey's day. In addition, women of the 1970s were more orgasmic than their earlier counterparts; 75 percent of single women reported having orgasms as compared with half of the young women in the Kinsey study.

A *Redbook* magazine survey of over 100,000 women was compiled in 1974 by Tavris and Sadd (1975). The sample includes women of all ages, educational backgrounds, religious beliefs, and political positions. The authors used a random sample of 18,000 married women, sometimes concentrating on only 2,278 cases for specific questions. They discovered, according to these women, that more frequent sex translated into a more satisfactory sex life. Of the women who were having sex eleven or more times a month, over 80 percent of them reported that their sex life was satisfactory. Women having sex less frequently were unhappy with their rate of intercourse and their overall sex life. Only 4 percent of the women thought that they were having too much sex. Orgasm, these women claimed, is purely a matter of partner technique: if husbands learn exactly how to stimulate wives, orgasm is easily achieved. Almost half of the wives said that they initiated sex half the time in the relationship. Seventy-five per cent said that they took an active part in sexual activity in bed, and only 13 percent could be classified as truly passive. The eroticism of these women, in contrast to what might be expected from women supposedly uninterested in sex, is quite surprising. For example, six wives out of ten had gone to a pornographic movie, seven out of ten liked erotic clothing, and 75 percent thought sex in various locations was exciting. Regardless of society's repression of female sexuality, modern women follow their natural tendency to enjoy sex. Although the *Redbook* study provides reliable data on a large sample of women and thus adds to a picture of American female sexuality, it's hampered by its significant limitation to married women. In addition, this report doesn't have comparative figures for males as does the Kinsey study. However, their information on female sexual behavior does describe a fairly high level of female sexual interest among married women who are under social sanctions to constrain their sexual behavior within a monogamous relationship.

Shere Hite's work on female sexuality, much publicized and pilloried, was less rigorous but nonetheless informative about more current standards of female sexuality in Western cultures (1976). Her sample of 3,019 women responded to a questionnaire. Like the Kinsey sample or the *Redbook* group, this sample can't be considered random, but the women are much more geographically diverse and of all socioeconomic classes, education, and ethnicity. In a sense, the Hite data are even more revealing because they represent freer and more elaborate answers from a more diverse population The striking feature of *The Hite Report* is the strong and quite varied language of the women themselves. Unhampered by scientific charts and levels of statistical significance, these words literally leap from the pages and define human female sexuality. The subjective response of the reader is that females are deeply concerned with their sexuality and the pleasure they receive.

This work is contrasted by several opinion polls conducted during the 1970s and 1980s. In 1988 and 1989 the National Opinion Research Council of the University of Chicago added questions about sex to their opinion surveys. They found that men and women were equally likely to be unfaithful to their monogamous partner (monogamy defined here as basically being with one partner for a year) during the previous year. Admittedly, "unfaithful" men had more different sexual partners than did "unfaithful" women, but this difference was due primarily to the classification of unmarried men. These unmarried men also had sex more often than did unmarried women, but married women and married men had had the same number of sexual encounters during the previous year. The author of this and other reports believe that men consistently overreport their sexual prowess and women underreport, a situation that undermines the credibility of the findings and perhaps adds a leveler to the disparate numbers between the sexes (Greeley, Michael, and Smith 1990, Smith 1991). Other surveys of sexual practice during the span of a year also questioned differences between men and women. In a survey of 1,016 men and 1,144 women conducted yearly from 1972 to 1989, only 10 percent more women than men had had only one sexual partner during the previous year, and men and women who had multiple partners seemed to have had almost an equal number (Wood 1990). But again, when women are without a steady partner, they reportedly have less sex than men without a relationship. But once hooked into a monogamous relationship, men and women seem to have sex just as often. Several studies questioned the so-called sexual revolution in America during the 1970s (Greeley, Michael, and Smith 1990, Klassen, Williams, and Levitt 1989, Smith, 1990). If people's opinions about what others should and shouldn't do are indicators, premarital sex in the postrevolution is more acceptable than it was before, but extramarital sex is just as reprehensible to the general public. But again, what people *say* everyone should do and what really happens behind closed doors are often different matters. More disturbing to our understanding of sexual behavior in our culture, these more scientific surveys are hampered by their methods: they ask respondents about their behavior for only a previous year, often lump married and unmarried individuals together, and ask more questions about opinions than about real sexual acts. This approach tells us little about adult behavior over a life course, through marriage or a long-term relationship. In addition, none of these more academic studies asked women what they thought about their sex lives, whether they were satisfied or not. In only one study conducted in 1970 were women and men asked if they found sex enjoyable. According to their data, 77 percent of the women and 94 percent of the men respondents said that they liked sex most of the time or every time (Klassen, Williams, and Levitt 1989). But no one knows whether the lack of female enjoyment is due to an innate dislike of the sexual act or to unsatisfactory sex with an inept partner. The more popular studies, in which women answer in depth about their feeling about sex, suggest the latter.

By far the largest survey on female sexuality was compiled by *Cosmopolitan* magazine (Wolfe 1980). The magazine sent out questionnaires to their readers, and no one was pretending that this group was a random sample. Writer Linda Wolfe, who compiled the data, points out with genuine surprise that the most amazing thing about the survey was the response: 106,000 women answered, the largest number of returns ever received by a magazine questionnaire. This in itself says something about the level of interest of modern women in their own, and others', sexual behavior. The article discusses the data for the whole sample and for a subset of women between the ages of eighteen and thirty-four who live in large cities—the "typical" *Cosmo* reader. They found that women of the 1980's had more lovers than had women in previous studies. Two-thirds of the women had had sex with five or more different men, and only 9 percent had experienced only one man. Fifty percent of the married *Cosmo* respondents said that they had had an affair while married. Only 10 percent of the women had never masturbated, and even most women in relationships continued to engage in self-sex. And 99 percent of the women in this survey had reached orgasm.

The Hite interviews, Tavris and Sadd, and Wolfe all report similar figures for extra marital affairs, which seem to occur more often for males than females. One then wonders, whom are these males having affairs with? Shouldn't the figures be equal unless a few single females are responsible for most of the affairs with married men? A survey of people in monogamous relationships, married and unmarried, conducted by the National Opinion Research Center reported only 1.5 percent had been unfaithful during the previous twelve months and that men and women were equally responsible for this number (Smith, 1991). One researcher discovered that when "opportunity" is factored into the disparate figures for male and female extramarital affairs, the data become similar for the two sexes. He suggests that males have more opportunities than females do (or they perceive more opportunities) and that if females had similar chances their rate of affairs would dramatically increase (Johnson 1970). This possibility is reinforced by the *Redbook* data in which working women were twice as likely as housewives to have extramarital affairs. Although the difference might be explained by a correlation between work and a heightened sex drive for these females, this explanation seems implausible. In fact, extramarital sex for part-time workers falls in between that for full-time workers and for full-time housewives. Figures for extramarital affairs have also steadily increased with each study, from Kinsey to Wolfe, an increase reflecting female sexual liberation in recent decades (Kinsey et al. 1953, Tavris and Sadd 1975, Wolfe 1980). Females usually cite sexual dissatisfaction or boredom within the marriage (more often than did males) as the motivation for looking elsewhere (Hunt 1969, Johnson 1970, Tavris and Sadd 1975). More interesting is the effect, or lack of effect, of these affairs on marriage. Most often, affairs occur during the first four years of marriage in the United States, and divorce results in only 1 to 2 percent of the cases. For 50 to 80 percent of the affairs, the spouses never even knew of the infidelity (Gebhard and Johnson 1979, Wolfe 1980). What these data tell us is that when women have extramarital affairs they don't intend to break up their marriages or have children with men who aren't their husbands (Essock-Vitale and McGuire 1985, Hunt 1969, Tavris and Sadd 1975). These affairs do in fact appear to be simple sexual liaisons for women seeking sexual pleasure outside the marriage.

Less tied to the hormonal fluctuations of a strict estrous cycle, human females may be motivated to engage in sexual behavior regardless of their potential for conception. In addition, orgasm isn't necessary for conception, and yet women can reach orgasm at any time given proper stimulation. Males, however, must have an orgasm to ejaculate male gametes. This major reproductive difference between males and females doesn't *necessarily* lead to a difference in sexual motivation. Human females, with their potential for orgasm, might also be interested in sex at any opportunity, regardless of the impact of that sexual behavior on reproductive success. In fact, the connection between sexual behavior and reproduction seems to be *less* for human females than for males. Although women must surely be careful about which male fathers their infants, this caution doesn't necessarily squash their sexuality. The level of parental investment required by females can be just as disconnected from most matings as these two levels can be separated in men, especially with reliable birth control. In a sense, women are that much freer than men to engage in sexual fun if they know when ovulation occurs and are careful during that time. Males, on the other hand, are always vulnerable to being caught in the web of parental investment, because unlike women they can't fully disconnect sex from reproduction. The information on human females demonstrates a certain insatiability for sex. This insatiability in females is so strong, some suggest, that males must restrict female sexuality (Goethals 1971), and most cultures do so in one way or another.

Female mate choices in all cultures are compromised by male power. Male power, in turn, is wielded to control female reproduction. As Blaffer Hrdy suggests, males wouldn't need to sequester females if females were sexually passive and they could be inseminated at male whim (1981). But when females are interested in sex, motivated to have sex, and males can't reliably predict ovulation, females are at risk for male domination. Almost universally, social and physical restrictions in the realm of sex are harsher on women than men (Gebhard and Johnson 1979). Women are often married off early or they are sequestered away from males until marriageable age. The only way for human males to ensure paternity is to restrict females. Part of that restriction may lie in convincing women that they are biologically less sexual and less intrigued by different sexual partners than are males.

Is There Female Choice among Modern Humans?

Because there are few strong morphological differences between male and female humans, we aren't likely to find the first kind of

female choice, the kind that affects the evolution of male characteristics ("Fisherian Choice" as outlined in Chapter 4). We humans aren't
particularly dimorphic in size or morphology. Most traits that appear
in males at puberty are primary sexual characteristics (enlarged testes,
lengthened penis) and thus have evolved by natural selection. Other
secondary sexual characteristics are shared by males and females alike
(axillary hair), and they too aren't candidates for sexually selected
traits. Even facial hair in males, which might be a sexually selected
character, exhibits extensive racial variation and therefore it's not
species specific. But there's good evidence for believing that, and
good evolutionary reasons why, human females might make the second kind of female choice ("Triverian Choice"), choice for their own
reproductive success.

When human females get the chance, do they make mate choices
and how exactly do they choose? David Buss, a scholar who looks at
human mating pattern from an evolutionary perspective, suggests
that human females need not attend to cues of male reproductive
value, better known as his ability to father children, because male
value reduces only slightly with age (1987). He joins those who see
the potential for paternal care of their infants as what women
"should" seek out. But females have difficulty in assessing potential
male care when males have no history of parenting. Resources are
easier to evaluate: females can assess status, goods, and financial
stability, all of which might aid in offspring longevity. In a study of
mate choices in the United States, Buss found that people in our
culture positively assort for age, race, religion, ethnic background,
socioeconomic status, and geographic location (1985). Using questionnaires, he asked men and women what they preferred in a mate,
not what they had actually chosen in the past. Some of these traits,
such as race and socioeconomic status, are of course highly correlated.
In addition, when couples chose for race, they also chose for similar
physical characteristics. Men and women differed slightly in what
they considered important in a mate; men ranked physical attractiveness higher than women did, and women preferred men with
financial stability. The female preference for male wealth and status
in our culture and others may actually be an example of truthful
advertising which works for men. For instance, wealthy men in the
United States sometimes have more children than poorer males and
hence higher reproductive success (Essock-Vitale 1984), although the

opposite data, with wealthy men having fewer children, have also been found (Vining 1986). But Buss's rankings of what's important to the two sexes also show the similarity between men and women; both share the same first several categories albeit in slightly different order. In evaluating marriage choices, we can focus on such things as race, religion, and political or economic values and try to figure out which might be high motivators for a marriage partnership, but this study does little to explain blind lust.

The situation in other cultures is more difficult to analyze, and it's no surprise that few studies have examined female choice. Most often, ethnographers write down what they observe and don't necessarily ask the same questions of all individuals. And then asking questions about women, sex, and marriage is a touchy business: no one really wants to divulge reasons for choosing a certain partner. In an extensive study of mate preferences in thirty-seven cultures, Buss looked for differences in preferences between the sexes (1989). Basing his assumptions on traditional sexual selection theory, he predicted that men would be interested in female reproductive potential—how many children a woman would be likely to produce—while women would be concerned with what a man could provide in resources to bring up those children. He did find some support for his hypotheses. In almost all of the cultures, men rated "looks" high on their agenda, and women were interested in earning potential. Women liked older men, who presumably have more resources, and men liked younger women, who are presumably fertile. But there are major flaws in both the data and their interpretation. The subjects were not chosen systematically, for example, and the information was gathered in all sorts of ways—a result to be expected of an attempt to launch a worldwide project. The supposed age difference, with men preferring younger women and women preferring older men, was really only a three-year age spread—not particularly meaningful in the real world. More important, Buss himself notes that what really emerges is not the differences between the sexes but what they share in common. Overwhelmingly, both men and women thought that a "kind-understanding" and "intelligent" person made the best mate, and these qualities were ranked higher than either looks or earning power (1989). He also suggests that a mutual preference for personality characteristics that are important for long-term relationships is a sign of a species-specific type of mate preference, one that outweighs any sex difference in choice. His work also shows that mate selection is important not just to women, but to men as well.

Although Buss's studies may tell us what people say they like, they tell us little about actual attraction and selection. In neither study were people questioned about the mates they had actually chosen, and we all know that what we say and what we do are often worlds apart. We still don't know what attracts one person to another.

When I decided to look at partner preference cross-culturally, I had to do what any other anthropologist does—I went through the available ethnographies and read the words of the ethnographers. This method is facilitated by a system set up years ago to aid those interested in cross-cultural practices. The system, called the Human Relations Area File (HRAF) includes original ethnographies coded by subject so that a researcher who is interested in, say, mating and marriage can go directly to that section and look at any number of cultures. Even better, the system is set up so that the scholar who wants to scan a sample of cultures from all over the globe and include every different type of culture, such as nomads and hunter-gatherers, can use a standard list of about two hundred which includes all sorts of subsistence types. It's called the Standard Cross-Cultural Sample (Murdock and White 1969). Using this material isn't as easy as it sounds. Sometimes more than one ethnographer writes about a culture, and one person's information can contradict another's. Also, most of the ethnographies are quite old, and a culture may now look different than it did when the ethnographer sat among the group. And because cultures evolve, change, and are influenced by contact with others, no one can check whether the words written by an ethnographer about a group were in fact true even at that time. But this is the essence of cultural anthropology, these stories of other peoples, and they are all we have to work with when we try to understand how universal a pattern of human behavior might be.

I was interested in determining whether women in most cultures have any say in choosing marriage partners and, once marriage occurs, whether women mate with men other than their husbands. Do female humans exert themselves in marriage and mating as a universal pattern? Because evolutionary theory predicts that women would be choosy and uninterested in sex with multiple partners and that men would be concerned with keeping the reproductive potential of wives to themselves, my search was an attempt to discover whether humans bear out these predictions (Small 1992a).

The Standard Sample has information on marriage and extramarital sex for 133 cultures outside of the Western sphere. Arranged marriages are, or were, found in 106 societies (80%), but surprisingly, only 23 (17%) of these cultures use arranged marriage as the exclusive way to form families. Only in three societies do males (grooms) have freedom to choose while females (brides) must accept male choice. Almost all groups that exhibit arranged marriages also accept marriages that are a surprise to the family and based on love. The main difference between arranged marriages and those not arranged is the amount of influence both partners might have on mate choice. In arranged marriages, families decide the partnership and don't consult either partner. When the bride and groom are allowed a voice in arrangements, men and women have an equal say in the arrangements of marriage in more than half of the groups (Whyte 1978). When couples have the freedom to make their own matches, families aren't involved until after the choices have been declared.

Most cultures also see a distinct difference between marriage and sexual matings, and it's marriage, rather than sexual interaction, that is important to all human cultures (Rosenblatt and Anderson 1981), a distinction that makes sense. Although sexual liaisons may result in offspring, marriages are long term and involve exchange of goods, land, and political alliances. This social importance of marriage is supported by the lack of the woman's voice when girls are married off in infancy and childhood even before they menstruate; these females don't have a voice because the proposed marriage is not a love match. But the literature on childhood engagements shows that at least one-quarter of the women are allowed to break the engagement when they become adults (Small 1992a). It should also be mentioned that young males, the grooms, are also subject to the same familial strong-arming.

Women have some ways to influence or escape arranged marriages. Because elopement and divorce occur in most of these cultures, a woman may not be helpless when her parents override her own preference. It's also possible that most females who are restricted to accept arranged marriage don't protest because they really don't care whom they marry. We don't know: the female voice was not in focus because the ethnographer didn't investigate the power of the bride's opinion, especially when that opinion may have been voiced behind closed doors. Also, because some wives in all societies are unfaithful, a woman need not look on an arranged marriage as the end of love. In addition, we can't assume that the interests of the families, the bride, and the groom are different (Irons 1979); if they are, women often protest, resist, and elope with their own choices (Buss 1989). Nor can we assume that the bride has no influence over her family during the courtship. Women often have a voice in arranged marriages; that is, they are able to reject offers in almost half of the societies that define their marriages as arranged. Thus "arranged" doesn't necessarily mean a woman is forced or coerced. According to evolutionary theory, brides "should" be interested in gaining mates of high status and resources, and certainly the family is interested in the same attributes. Thus arrangements, even when forced, may not be particularly distasteful to women.

Although those of us in western culture usually squirm at the thought that someone else might determine whom we should marry, we can't assume that an arranged marriage is more unsatisfying sexually or emotionally than a nonarranged one. Before an arranged marriage, there's always room for a strong fantasy life that might heighten attraction and attachment (Rosenblatt and Anderson 1981). In addition, ceremonial trappings serve to elevate the relationship past a simple arrangement and increase the excitement of the moment. At the same time, there's less pressure for compatibility and fewer expectations; because no one expects this marriage to be wonderful, partners might settle for less. Freedom of choice, on the other hand, usually leads to focusing on impractical and often transitory emotions such as love and traits such as sexual attraction. Then too, "freedom" is relative term, because opportunities for "falling in love" are bounded by cultural opportunities that usually throw appropriate, culturally sanctioned individuals together. All cultures also have rules of exogamy, marrying outside one's group, which dictate social and familial norms and define even the freest choices: you are likely to marry someone of your own race, who often looks much like you and has the same religion and political beliefs because you have the same friends and frequent the same places (Diamond 1992). And even in free marriages, parental approval is almost always sought and hoped for anyway.

When women in other cultures do have a chance to filter among the available males and decide on one, whom do they choose? No one really knows. There's no information about what characteristics might be important. We do know that, cross-culturally, high status men in more traditional societies often have more wives and more children than do low-status men, and that wives often choose to be in polygynous marriages (Betzig 1988, Borgerhoff Mulder 1990, Dickemann 1979, Flinn 1986, Irons 1979, 1980, Turke and Betzig 1985). In addition, high-status women also have higher fertility and a higher rate of survival for their infants (Irons 1979, 1980). But whether or not women value and choose high-status males only because of their potential is hard to say, because no one has questioned them about their or their families' initial motives. The question of what females prefer, and how that preference affects current marriage practices among humans worldwide, remains an open question.

Les Liaisons Dangereuses

Once married, women and men are usually expected to mate within the union. Although many societies have polygynous marriages, the majority of human unions are monogamous. Are we naturally monogamous? Is the monogamous state of marriage a reflection of our sexual preference or something imposed to help with children and extended families? Evidence I've cited earlier suggests that although we may bond over the long term in pairs our mating behavior is anything but monogamous. Also, in the United States and Great Britain a few genetic studies have shown that about 10 percent of infants born in at least a few hospitals couldn't have been conceived with the man recorded as the father (Diamond 1992), and of course most copulations don't result in conceptions, for a number of reasons.

Again I used the HRAF to look at extramarital affairs in other cultures. I hypothesized that if human females and males marry primarily to raise children and make family alliances, we might expect a conflict at the level of mating. We can be considered naturally preprogrammed monogamous beings only if our pair-bonded marriages are in fact just that—an exclusive sexual relationship. In the 133 cross-cultural societies I evaluated, at least some women in all cultures engage in extramarital affairs. Several anthropologists, using the same cross-cultural technique, also found that extramarital affairs by men and women are present in most cultures (Broude and Greene 1976, Whyte 1978). Broude and Green, for example, report that males in 80 percent of the cultures and females in 73 percent of the cultures engage in extramarital liaisons.

Most societies disapprove of extramarital philandering, both by men and women. Every so often, women in particular cultures are "allowed," even encouraged, to participate in sexual behavior with men who aren't their husbands, but this infidelity is usually a sign that female sexuality is controlled by husbands. For example, wife exchange (or swapping) establishes and reinforces reciprocal altruism and alliances among men, and when such an exchange occurs it's usually perpetrated by husbands. Fraternal polyandry, in which brothers share a woman, potentially increases a husband's inclusive reproductive fitness, although this marriage system is rare (Broude 1980).

In only a third of the societies in the Standard Sample are extramarital affairs really acceptable, or even expected as a possibility, for women (Small 1992a). But interestingly enough, mention of male infidelity never occurs without mention of female infidelity. In other words, both of the sexes have affairs. Punishment for women who stray ranges from malicious gossip and social ostracism to physical abuse and death. In some groups, women were, and still are, mutilated to keep them from engaging in sex with anyone other than their husbands. Infibrulation, in which the labia are sewn together, is still practiced in at least twenty-three societies (Lightfoot-Klein 1989). The stitches are undone for childbirth or for insemination for the next conception, but kept closed the rest of the time. Women are also subject to clitorectomy, in which the clitoris is removed. Although this practice is now a tradition and thus often accepted by women as well as men, its origins must have been to control female sexual pleasure and, as a consequence, paternity. The men must have seen women as wild sexual beings who would mate with anyone if they weren't "fixed."

The "double standard" is really a statement about the power to control (or attempt to control) rather than about differences in male and female sexuality (Rosenblatt and Anderson 1981). Extramarital affairs are allowed, or simply ignored, for men while the punishment for women is often severe. But as one anthropologist points out, just because women are punished more often, it doesn't necessarily follow that women are less interested in extramarital affairs (Broude 1980). On the contrary, women are probably restricted *because* men are frightened of the sexual and reproductive potential of women. The more important point is that women continue to have extramarital affairs, even under the threat of harsh punishment. This fact underscores the strong human female sex drive.

Female Mate Choice in Conclusion

Certain inherent flexibilities in human mating strategies reflect physiological and social abilities and constraints. Men are able to fertilize several women, but their powers of insemination aren't exhaustive. Women have a limited number of eggs, mate throughout their cycles and nonseasonally, and ovulation is somewhat concealed. Because of these differences, men and women may have dynamically different strategies in the long and short term when they choose partners for marriage or for brief sexual liaisons (Buss and Schmitt in press). But similar constraints on both men and women delimit human reproduction and mold mating strategies into a compromise. Lacking flamboyant advertisement of female fertility, neither women nor men are able to pinpoint ovulation and control reproduction. Human females are also restrained in their reproductive output by highly dependent offspring. Males too are constrained by the slow development of the human infant, that packet of reproductive success which evolution has pushed to the extreme of dependency. Although the details of human sexuality clearly show differences in male and female sexual response and actions, the similarities are also striking. Women respond to arousal as quickly as men, women are just as interested in sexual satisfaction as men, and given the proper partner, women will enjoy sex as much as a man. Cross-culturally, women engage in extramarital affairs almost as often as men, or they wish they could. And most societies disapprove of such behavior. Both men and women play the mate choice game, presumably to improve reproductive success, and both must bow to the wishes of families and operate under the rules of society.

If we're so monogamous, why is there so much philandering before, during, and after marriage? Even in the face of the dire consequences of AIDS, why do women and men seek multiple partners? And more important, if women are supposed to be choosy, why are they, in particular, choosing to mate outside the bond? It may be that a woman is executing an evolutionary strategy when she commits adultery. Perhaps she's just trying to improve her chance of conceiving when ovulation is not particularly detectable, or maybe she's seeking a biological father who is perhaps better than the man she's committed to. The result, in any case, is an inherent conflict between the levels of marriage and sex . Marriage, a particularly human expression of a pair-bonded social system, probably arose to accommodate dependent human offspring. Culturally, marriage is also an economic and political union within and across communities. But the conflict arises because of our additional inherent sexual nature, one that disconnects sexual activity from parenting. Thus our social system doesn't correspond exactly with our mating system, and we humans live in constant tension over how to live our sexual lives.

In this sense, no choice of a mate and no choice of a sexual act is really a free choice. Each is driven by the biological urge to enjoy the pleasures of the body. Sexual behavior also happens to result in offspring. The passage of genes, the establishment of alliances, and the need for economic security are concerns not just of the mating pair, and by this I mean men and women equally, but also of the larger familial gene pool.

Female Nature

We can't escape our biological heritage. We share certain attributes with other female primates—large body size compared with other mammals, big brains and social smarts, a tracking mechanism for kinship and friendship, varied periods of sexual interest, and a drive for sexual pleasure. On top of this, our infants demand great attention and investment. Nonhuman primates, with their exciting array of mating strategies, tell us that humans are just one more type of primate looking for ways to be sexually satisfied and at the same time pass on genes to the next generation. Our compromise has been the evolution of a human pair-bond that isn't necessarily exclusive.

In knowing the biological basis of our sexual behavior we aren't strapped into a strait jacket of biological determinism. Instead, we are more liberated from what society tells us we "should" be like. No matter what our culture tells us, no matter the changes that push or pull our sexual nature, the basic core remains the same. We can easily turn to our female primate cousins and show that, like them, we're strategizing female creatures, concerned with our sexual and reproductive lives. The demands of parenting have selected for a particular social system and have given men more power over our reproductive decisions than they should have. We should, perhaps, follow the example of our bonobo sisters, who have parlayed their sexual nature into equality with bonobo males.

For females searching for their roots, the road map of evolutionary history discussed in this book, with all its alternate routes, junctions, dead ends, and forks, is an essential ingredient to understanding who we are. Just as human nature is part of the larger picture of the animal kingdom, so too is female nature connected to all creatures female. So on your next visit to the zoo, look inside the cage of the female chimpanzee while she presents her swelling to a male; watch the female lemurs groom one another as they make an alliance; catch a female baboon running from a would-be consort male. Feel the strands of DNA unravel and connect you and those brown eyes that stare back though the bars of the cage. And acknowledge that primateness, and sisterhood, extends beyond one family, one community, one race, or one species.

References

Adams, D.B., A.R. Gould, and A.D. Burt. 1978. Rise in female-initiated sexual activity at ovulation and its suppression by oral contraceptives. *New England Journal of Medicine* 299: 1145–1150.

Akers, J., and C. Conaway. 1979. Female homosexual behavior in *Macaca mulatta*. *Archives of Sexual Behavior* 8: 63–80.

Alexander, R.D. 1990. How did humans evolve? *University of Michigan Special Publications* 1: 1–38.

Alexander, R.D., and K.M. Noonan. 1979. Concealment of ovulation, parental care, and human social evolution. In *Evolutionary Biology and Human Social Behavior*, ed. N.A. Chagnon and W.G. Irons, pp. 436–453. North Scituate, Mass.: Duxbury Press.

Altmann, J.A. 1980. *Baboon Mothers and Infants*. Cambridge: Harvard University Press.

Altmann, S.A., and J. Altmann. 1970. *Baboon Ecology*. Chicago: University of Chicago Press.

Andelman, S.J. 1987. Evolution of concealed ovulation in vervet monkeys (*Cercopithecus aethiops*). *American Naturalist* 129: 785–799.

Anderson, A. 1992. The evolution of sexes. *Science* 257: 324–326.

Andersson, M. 1982. Female choice selects for extreme tail length in a widowbird. *Nature* 299: 183–186.

Andersson, M. 1986. Evolution of condition-dependent sex ornaments and mating preferences; Sexual selection based on viability difference. *Evolution* 40: 804–816.

Andersson, M., and J.W. Bradbury. 1987. Introduction. In *Sexual Selection: Testing the Alternatives*, ed. J.W. Bradbury and M.B. Andersson, pp 1–8. New York: John Wiley.

Bachmann, C., and H. Kummer. 1980. Male assessment of female choice in Hamadryas baboons. *Behavioral Ecology and Sociobiology* 6: 315–321.

Basolo, A. 1990. Female preference for male sword length in the green sword-tail, *Xyphophorus helleri*. *Animal Behaviour* 40: 332–338.

Basolo, A. 1991. Female preference predates the evolution of the sword in swordtail fish. *Science* 250: 808–810.

Baxter, M.J., and L.M. Fedigan. 1979. Grooming and consort partner selection in a troop of Japanese monkeys. *Archives of Sexual Behavior* 8: 445–458.

Beach, F. 1976. Sexual attractivity, proceptivity, and receptivity in female mammals. *Archives of Sexual Behavior* 14: 529–537.

Bearder, S.K. 1987. Lorises, bushbabies, and tarsiers: Diverse societies in solitary foragers. In *Primate Societies*, ed. B. B. Smuts, D. L. Cheney, R.M. Seyfarth, R. Wrangham and T.T. Strusaker, pp. 11–24. Chicago: University of Chicago Press.

Beckwith, C. 1983. Niger's Wodaabe: People of the taboos. *National Geographic* 64: 483–509.

Bell, G. 1982. *The Masterpiece of Nature*. Berkeley: University of California Press.

Benshoff, L., and R. Thornhill. 1979. The evolution of monogamy and concealed ovulation in humans. *Journal of Social and Biological Structures* 2: 95–106.

Berard, J.D., P. Nurnberg, J.T. Epplen, and J. Schmidtke. Reproductive success of male rhesus macaques as revealed by oligonucleotide fingerprints. Unpublished manuscript.

Berenstain, L., P.S. Rodman, and D.G. Smith. 1981. Social relationships between fathers and offspring in a captive group of rhesus monkeys (*Macaca mulatta*). *Animal Behaviour* 29: 1057–1063.

Berkovitch, F. 1991. Mate selection, consortship formation, and reproductive tactics in adult female savannah baboons. *Primates* 32: 437–452.

Berman, C.M. 1984. Variation in mother-infant relationships: Traditional and nontraditional factors. In *Female Primates: Studies by Women Primatologists*, ed. M.F. Small, pp. 17–36. New York: Alan R. Liss.

Bernstein, H., F.A. Hopf, and R.E. Michod. 1988. Is meiotic recombination an adaptation for repairing DNA, producing genetic variation, or both? In *The Evolution of Sex*, ed. R.E. Michod and B.R. Levin, pp. 139–160. Sunderland, Mass.: Sinauer Associates.

Betzig, L. 1988. Mating and parenting in Darwinian perspective. In *Human Reproductive Behavior*, ed. L. Betzig, M. Borgerhoff Mulder, and P. Turke, pp. 3–20. Cambridge: Cambridge University Press.

Bielert, C., J.A. Czaja, S. Eisle, G. Scheffler, J.A. Robinson, and R.W. Goy. 1976. Mating in the rhesus monkey (*Macaca mulatta*) after conception and its relationship to oestradiol and progesterone levels throughout pregnancy. *Journal of Reproduction and Fertility* 46: 179–187.

Blaffer Hrdy, S. 1977. *The Langurs of Abu*. Cambridge: Harvard University Press.

Blaffer Hrdy, S. 1979. The evolution of human sexuality: The latest word and the last. *Quarterly Review of Biology* 54: 309–314.

Blaffer Hrdy, S. 1981a. *The Woman That Never Evolved*. Cambridge: Harvard University Press.

Blaffer Hrdy, S. 1981b. "Nepotists" and "altruists": The behavior of old females among macaque and langur monkeys. In *Other Ways of Growing Old*, ed. P. Amoss and S. Harrell, pp. 119–146. Palo Alto: Stanford University Press.

Blaffer Hrdy, S. 1983. Heat loss. *Science '83* October: 73–78.

Blaffer Hrdy, S., and D.B. Hrdy. 1976. Hierarchical relationships among female Hanuman langurs (*Presbytis entellus*). *Science* 193: 913–915.

Blaffer Hrdy, S., and P.L. Whitten. 1987. Patterning of sexual activity. In *Primate Societies*, ed. B.B. Smuts, D.L. Cheney, R.M. Seyfarth, R. Wrangham and T.T. Strusaker, pp. 370–384. Chicago: University of Chicago Press.

Blaffer Hrdy, S.B., and G.C. Williams. 1983. Behavioral biology and the double standard. In *Social Behavior of Female Vertebrates*, ed. S.K. Wasser, pp. 3–17. New York: Academic Press.

Boak, C.R. 1986. A method for testing adaptive hypotheses of mate choice. *American Naturalist* 127: 654–666.

Boinski, S. 1987. Mating patterns in squirrel monkeys (*Saimiri oerstedii*). *Behavioral Ecology and Sociobiology* 21: 13–21.

Borgerhoff Mulder, M. 1990. Kipsigis women's preference for wealthy men: Evidence for female choice in mammals? *Behavioral Ecology and Sociobiology* 27: 255–264.

Broude, G.J. 1980. Extra-marital sex norms in cross-cultural perspective. *Behavioral Science Research* 15: 181–218.

Broude, G.J., and S.J. Greene. 1976. Cross-cultural codes on twenty sexual attitudes and practices. *Ethnology* 60: 409–429.

Brown Blackwell, A. 1875. *The Sexes through Nature*. New York: G.P. Putnam's.

Burley, N. 1979. The evolution of concealed ovulation. *American Naturalist* 114: 835–858.

Burt Gamble, E. 1894. *The Sexes in Science and History: An Inquiry into the Dogma of Women's Inferiority to Man*. New York: G.P. Putnam's.

Burton, F.B. 1971. Sexual climax in female *Macaca mulatta*. *Proceeding of the Third International Congress of Primatology* 3: 180–191.

Buss, D.M. 1985. Human mate selection. *American Scientist* 73: 47–51.

Buss, D.M. 1987. Sex differences in human mate choice criteria: An evolutionary perspective. In *Sociobiology and Psychology: Ideas, Issues, and Applications*, ed. C. Crawford, M. Smith, and D. Krebs, pp. 335–351. Hillsdale, N.J.: Lawrence Erlbaum Associates.

Buss, D.M. 1989. Sex differences in human mate preferences: Evolutionary hypotheses tested in 37 cultures. *Behavioral and Brain Sciences* 12: 1–49.

Buss, D.M., and D.P. Schmitt. Sexual strategies theory: A contextual, evolutionary analysis of human mating. *Psychological Review*. In press.

Busse, C., and W.J. Hamilton. 1981. Infant carrying by male chacma baboons. *Science* 212: 1281–1283.

Caine, N.G., and G. Mitchell. 1979. A review of play in the genus *Macaca*: Social correlates. *Primates* 20: 535–546.

Caine, N.G., and G. Mitchell. 1980. Species differences in interest shown in infants by juvenile female macaques (*Macaca radiata* and *Macaca mulatta*). *International Journal of Primatology* 1: 323–332.

Carpenter, C.R. 1942. Sexual behavior of free ranging rhesus monkeys (*Macaca mulatta*). *Journal of Comparative Psychology* 33: 113–142.

Chapais, B. 1986. Why do adult male and female rhesus monkeys affiliate during birth season? In *The Cayo Santiago Macaques: History, Behavior, and Biology*, ed. R.G. Rawlins and M.J. Kessler, pp. 173–200. Albany: State University of New York Press.

Cheney, D.L. 1978. Interactions of immature male and female baboons with adult females. *Animal Behaviour* 26: 389–408.

Cheney, D.L., and R.M. Seyfarth. 1990. *How Monkeys See the World*. Chicago: University of Chicago Press.

Cheney, D.L., R.M. Seyfarth, S.J. Andelman, and P.C. Lee. 1988. Reproductive success in vervet monkeys. In *Reproductive Success*, ed. T. Clutton-Brock, pp. 384–402. Chicago: University of Chicago Press.

Cheney, D.L., R.M. Seyfarth, and B.B. Smuts. 1986. Social intelligence and the evolution of the primate brain. *Science* 243: 1361–1366.

Chevalier-Skolnikoff, S. 1974. Male-female, female-female, and male-male sexual behavior in the stumptail monkey, with special attention to the female orgasm. *Archives of Sexual Behavior* 3: 95–116.

Chevalier-Skolnikoff, S. 1975. Heterosexual copulatory patterns in stumptail macaques (*Macaca arctoides*) and in other macaque species. *Archives of Sexual Behavior* 4: 119–220.

Chevalier-Skolnikoff, S. 1976. Homosexual behavior in a laboratory group of stumptail monkeys (*Macaca arctoides*). *Archives of Sexual Behavior* 5: 511–528.

Clutton-Brock, T., and P.H. Harvey. 1976. Evolutionary rules and primate societies. In *Growing Points in Ethology*, ed. P.P.G. Bateson and R.A. Hinde, pp. 195–237. Cambridge: Cambridge University Press.

Cohen, J., and D. McNaughton. 1974. Spermatozoa: The probable selection of a small population by the genital tract of the female rabbit. *Journal of Reproduction and Fertility* 39: 297–310.

Conaway, C.H., and C.B. Koford. 1965. Estrous cycles and mating behavior in a free-ranging band of rhesus monkeys. *Journal of Mammology* 45: 577–588.

Cords, M. 1984. Mating patterns and social structure in redtail monkeys (*Cercopithecus ascanius*). *Zeitschrift für Tierpsychologie* 64: 313–329.

Cords, M. 1987. Forest guenons and patas monkeys: Male-male competition in one-male groups. In *Primate Societies*, ed. B.B. Smuts, D.L. Cheney, R.M. Seyfarth, R. Wrangham and T.T. Strusaker, pp. 98–120. Chicago: University of Chicago Press.

Crockett, C.M. 1984. Emigration of red howler monkeys and the case for female competition. In *Female Primates: Studies by Women Primatologists*, ed. M.F. Small, pp. 159–173. New York: Alan R. Liss.

Crockett, C.M. 1985. Population studies of red howler monkeys (*Alouatta seniculus*). *National Geographic Research* 1: 264–273.

Curie-Cohen, M., D. Yoshihara, L. Luttrell, K. Bentorado, J. MacCluer, and W.H. Stone. 1983. The effects of dominance on mating behavior and paternity in a captive troop of rhesus monkeys (*Macaca mulatta*). *American Journal of Primatology* 5: 127–138.

Dahlberg, F. 1981. *Woman the Gatherer*. New Haven: Yale University Press.

Daly, M, and M. Wilson. 1978. *Sex, Evolution, and Behavior*. North Scituate: Duxbury Press.

Darwin, C. 1859. *The Origin of Species by Means of Natural Selection*. London: John Murray. Facsimile Avenel Edition. New York: Crown, 1979.

Darwin, C. 1871. *The Descent of Man, and Selection in Relation to Sex*. London: John Murray. Facsimile Princeton University Press Edition. Princeton: Princeton University Press, 1981.

Deag, J.M., and J.H. Crook. 1971. Social behavior and "agonistic buffering" in the wild Barbary macaque (*Macaca sylvanus* L.). *Folia Primatologica* 15: 183–200.

deRuiter, J.R., W. Scheffran, G.J.J.M. Trommelen, A.G. Uiterlinden, R.D. Martin, and J.A.R.A.M. vanHooff. 1992. Male social rank and reproductive success in wild long-tailed macaques, paternity exclusion by blood protein analysis and DNA fingerprinting. In *Paternity in Primates: Genetic Tests and Theories*, ed. R.D. Martin and A.F. Dixson, pp. 175–191. Basel: Karger.

de Waal, F. 1987. Tension regulation and nonrepoductive functions in captive bonobos (*Pan paniscus*). *National Geographic Research* 3: 318–335.

de Waal, F. 1990. *Peacemaking among Primates*. Cambridge: Harvard University Press.

Dewsbury, D.A. 1982. Ejaculate cost and male choice. *American Naturalist* 119: 601–610.

Diamond, J. 1992. *The Third Chimpanzee*. New York: Harper Collins.

Dickemann, M. 1979. The ecology of mating systems in hypergynous dowery societies. *Social Science Information* 18: 163–195.

Dittus, W. 1977. The social regulation of population density and age-sex distribution in the toque monkey. *Behaviour* 63: 281–322.

Dittus, W. 1979. The evolution of behavior regulating density and age-specific sex ratios in a primate population. *Behaviour* 69: 265–302.

Dixon, A.F. 1978. Observation on the evolution of the genitals and copulatory behavior in male primates. *Journal of Zoology* 213: 423–443.

Dunbar, R.I.M. 1984. *Reproductive Decisions*. Princeton: Princeton University Press.

Dunbar, R.I.M., and P. Dunbar. 1988. Maternal time budgets of gelada baboons. *Animal Behaviour* 36: 970–980.

Dunbar, R.I.M., and M. Sharman. 1983. Female competition for access to males affects birth rate in baboons. *Behavioral Ecology and Sociobiology* 13: 157–159.

Eaton, G.G. 1973. Social and endocrine determinants of sexual behavior of simian and prosimian females. *Symposium of the IVth International Congress of Primatology* 2: 20–35.

Eberhard, W.G. 1990. Animal genitalia and female choice. *American Scientist* 78: 134–141.

Enomoto, T. 1974. The sexual behavior of Japanese monkeys. *Journal of Human Evolution* 3: 351–372.

Enomoto, T. 1978. On social preference in sexual behavior of Japanese monkeys (*Macaca fuscata*). *Journal of Human Evolution* 7: 283–293.

Essock-Vitale, S.M. 1984. The reproductive success of wealthy Americans. *Ethology and Sociobiology* 5: 45–49.

Essock-Vitale, S.M., and M.T. McGuire. 1985. Woman's lives viewed from an evolutionary perspective. I. Sexual histories, reproductive success and demographic characteristics of a random sample of American women. *Ethology* 6: 137–154.

Fairbanks, L.A., and M.T. McGuire. 1985. Relationships of vervet monkeys with sons and daughters from one through three years of age. *Animal Behaviour* 33: 40–50.

Faludi, S. 1991. *Backlash*. New York: Crown.

Farr, J.A. 1977. Male rarity or novelty, female choice behavior, and sexual selection in the guppy, *Poecilea reticulata*, PETERS. *Evolution* 31: 162–168.

Fedigan, L.M. 1983. Dominance and reproductive success in primates. *Yearbook of Physical Anthropology* 26: 91–129.

Fedigan, L.M. 1986. The changing role of women in models of human evolution. *Annual Review of Anthropology* 15: 25–66.

Fedigan, L.M., and H. Gouzoules. 1978. The consort relationship in a troop of Japanese monkeys. In *Recent Advances in Primatology*, ed. D.J. Chivers and J. Herbert, pp. 493–495. New York: Academic Press.

Fisher, H. 1982. *The Sex Contract*. New York: William Morrow.

Fisher, R.A. 1930. *The Genetical Theory of Evolution*. New York: Dover.

Fleagle, J.G. 1988. *Primate Adaptation and Evolution*. New York: Academic Press.

Flinn, M.V. 1986. Correlates of reproductive success in a Caribbean village. *Human Ecology* 14: 225–243.

Foerg, R. 1982. Reproductive behavior in *Varecia variegata*. *Folia Primatologica* 38: 108–121.

Ford, C.A., and F.A. Beach. 1951. *Patterns of Sexual Behavior*. New York: Harper.

Fossey, D. 1983. *Gorillas in the Mist*. Boston: Houghton Mifflin.

Frayser, S.G. 1985. *Varieties of Sexual Experience: An Anthropological Perspective on Human Sexuality*. New Haven: HRAF Press.

Freedman, D.H. 1992. The aggressive egg. *Discover* 13: 60–65.

Garbers, D., and M. Eisenbach. 1991. Sperm attraction to a follicular factor(s) correlates with human egg fertilizability. *Proceedings of the National Academy of Science* 88: 2840–2844.

Gebhard, P. 1971. Human sexual behavior. In *Human Sexual Behavior: Variation across the Ethnographic Spectrum*, ed. D.S. Marshall and R.C. Suggs, pp. 206–217. New York: Basic Books.

Gebhard, P.H., and A.B. Johnson. 1979. *The Kinsey Data: Marginal Tabulations of the 1938–1963 Interviews Conducted by the Institute for Sex Research*. Philadelphia: W. H. Saunders.

Gibbons, A. 1992. More diverse than a barrel of monkeys. *Science* 255: 287–288.

Glander, K.E. 1984. Group composition in mantled howling monkeys. *American Journal of Physical Anthropology* 63: 163.

Glesener, R.R., and D. Tilman. 1978. Sexuality and the components of environmental uncertainty. *American Naturalist* 112: 659–673.

Goethals, G.W. 1971. Factors affecting permission and nonpermission rules regarding premarital sex. In *Sociology of Sex*, ed. J.M. Henslin, pp. 9–26. New York: Appleton-Century Crofts.

Goldfoot, D.A., H. Westerborg-van Loon, W. Groenveld, and A. Koos Slob. 1980. Behavioral and physiological evidence of sexual climax in the female stumptail macaque (*Macaca arctoides*). *Science* 208: 1477–1479.

Goodall, J. 1971. *In the Shadow of Man*. London: Collins.

Goodall, J. 1986. *The Chimpanzees of Gombe*. Cambridge: Harvard University Press.

Goodman, M. 1963. Man's place in the phylogeny of the primates as reflected by serum proteins. In *Classification and Human Evolution*, ed. S.L. Washburn, pp. 204–235. Chicago: Aldine.

Gould, S.J. 1987. Freudian slip. *Natural History* April: 15–21.

Gouzoules, H. 1980. The alpha female: Observations on captive pigtail macaques. *Folia Primatologica* 33: 46–56.

Gouzoules, H., and R.W. Goy. 1983. Physiological and social influences on mounting behavior of troop living female monkeys (*Macaca fuscata*). *American Journal of Primatology* 5: 39–49.

Goy, R., and R.J. Resko. 1972. Gonadal hormones and behavior of normal and pseudohermaphroditic nonhuman female primates. *Recent Progress in Hormonal Research* 28: 707–733.

Greeley, A.M., R.T. Michael, and T.W. Smith. 1990. Americans and their sexual partners: A most monogamous people. *Society* 27: 36–42.

Gust, D.A., and T.P. Gordon. 1991. Male age and reproductive behavior in sooty mangabeys (*Cercocebus torquatus atys*). *Animal Behaviour* 41: 277–283.

Halliday, T.R. 1983. The study of mate choice. In *Mate Choice*, ed. P. Bateson, pp. 3–32. Cambridge: Cambridge University Press.

Halliday, T., and S.J. Arnold. 1987. Multiple matings by females: A perspective from quantitative genetics. *Animal Behaviour* 35: 939–941.

Hamilton, W.D. 1964. The genetical evolution of social behavior. *Journal of Theoretical Biology* 7: 1–51.

Hamilton, W.D. 1984. Significance of paternal investment by primates to the evolution of male-female associations. In *Primate Paternalism*, ed. D.M. Taub, pp. 309–335. New York: Van Nostrand Reinhold.

Hamilton, W.D. 1988. Sex and disease. In *The Evolution of Sex*, ed. R. Bellig and G. Stevens, pp. 65–95. San Francisco: Harper and Row.

Hamilton, W.D., R. Axelrod, and R. Tanese. 1990. Sexual reproduction as an adaptation to resist parasites: A review. *Proceedings of the National Academy of Science* 87: 3566–3573.

Hamilton, W.D., and M. Zuk. 1982. Heritable true fitness and bright birds: A role for parasites? *Science* 218: 309–323.

Hamilton, W.J., and P.C. Arrowood. 1978. Copulatory vocalization of chacma baboons (*Papio ursinus*), gibbons (*Hylobates hoolock*), and humans. *Science* 200: 1405–1409.

Hanby, J.P., L.T. Robertson, and C.H. Phoenix. 1971. The sexual behavior of a confined troop of Japanese macaques. *Folia Primatologica* 16: 123–143.

Harcourt, A.H. 1979. Social relationships between adult male and female mountain gorillas. *Animal Behaviour* 27: 325–342.

Harding, R.S.O., and D.K. Olson. 1986. Patterns of mating among male patas monkeys (*Erythrocebus patas*) in Kenya. *American Journal of Primatology* 11: 343–358.

Harlow, H.F., and M.K. Harlow. 1965. The affectional systems. In *Behavior of Nonhuman Primates*, ed. A.M. Schrier, H.F. Harlow and F. Stollnitz, pp. 287–334. New York: Academic Press.

Hasegawa, T., and M. Hiraiwa-Hasegawa. 1990. Sperm competition and mating behavior. In *The Chimpanzees of the Mahale Mountains: Sexual and Life History Strategies*, ed. T. Nishida, pp. 115–132. Tokyo: University of Tokyo Press.

Hauser, M.D. 1988. Invention and social transmissions: New data from wild vervet monkeys. In *Machiavellian Intelligence: Social Expertise and the Evolution of Intellect in Monkeys, Apes, and Humans*, ed. R.W. Byrne and A. Whiten, pp. 327–343. Oxford: Clarendon Press.

Hauser, M. 1990. Do chimpanzee copulation calls incite male-male competition? *Animal Behaviour* 39: 596–597.

Hausfater, G. 1975. Dominance and reproduction in baboons. *Contributions to Primatolology* 7: 1–150.

Hausfater, G., and S., Blaffer Hrdy. 1984. *Infanticide: Comparative and Evolutionary Perspectives*. New York: Aldine.

Heisler, L, M.B. Andersson, S.J. Arnold, C.R. Boake, G. Borgia, M. Hausfater, M. Kirkpatrick, R. Lande, J. Maynard Smith, P. O'Donald, A.R. Thornhill, and F.J. Weissing. 1987. The evolution of mating preferences and sexually selected traits. In *Sexual Selection: Testing the Alternatives*, ed. J.W. Bradbury and M.B. Andersson, pp. 91–118. New York: John Wiley.

Hendrickx, A.G., and D.G. Kraemer. 1969. Observations on the menstrual cycle, optimal mating time, and pre-implantation embryos of the baboon, *Papio anubis* and *Papio cynocephalus*. *Journal of Reproduction and Fertility Suppliment* 6: 119–128.

Hershkovitz, P. 1977. *Living New World Monkeys (Platyrrhini)*. Vol. 1. Chicago: University of Chicago Press.

Hite, S. 1976. *The Hite Report*. New York: Dell.

Hollingsworth, M.J., and J. Maynard Smith. 1955. The effects of inbreeding on rate of development and on fertility in *Drosophila subobscura*. *Journal of Genetics* 53: 295.

Houde, A.E., and J.A. Endler. 1990. Correlated evolution of female mating preferences and male color patterns in the guppy *Poecilia retculata*. *Science* 248: 1405–1408.

Hubbard, R. 1979. Have only men evolved? In *Women Look at Biology Looking at Women*, ed. R. Hubbard, M.S. Henifin and B. Fried, pp. 7–35. Boston: G.K. Hall.

Huffman, M.A. 1987. Consort intrusion and female mate choice in Japanese macaques (*Macaca fuscata*). *Ethology* 75: 221–234.

Huffman, M.A. 1991a. History of the Arashiyama Japanese macaques in Koyoto, Japan. In *The Monkeys of Arashiyama: Thirty-five Years of Research on Japan and the West*, ed. L.M. Fedigan and P.J. Asquith, pp. 21–53. Albany: State of New York University Press.

Huffman, M.A. 1991b. Mate selection and partner preference in female Japanese macaques. In *The Monkeys of Arashiyama: Thirty-five Years of Research on Japan and the West*, ed. L.M. Fedigan and P.J. Asquith, pp. 101–122. Albany: State of New York University Press.

Huffman, M.A. 1992. Influences of female partner preference on potential reproductive outcome in Japanese macaques. *Folia Primatologica* 59: 77–88.

Hunt, M. 1969. *The Affair*. New York: World.

Hunt, M. 1974. *Sexual Behavior in the 1970's*. Chicago: Playboy Press.

Hurst, L., and W.D. Hamilton. 1992. Cytoplasmic fusion and the nature of the sexes. *Proceedings of the Royal Society of London* 247 (Series B): 189–194.

Huxley, J.S. 1938a. The present standing on the theory of sexual selection. In *Evolution: Essays on Aspects of Evolutionary Biology*, ed. G. R. deBeer. Oxford: Clarendon Press.

Huxley, J.S. 1938b. Darwin's theory of sexual selection and the data subsumed by it, in the light of recent research. *American Naturalist* 72: 416–433.

Imanishi, K., and S.A. Altmann. 1965. *Japanese Monkeys*. Atlanta: Emory University Press.

Irons, W.G. 1979. Cultural and biological success. In *Evolutionary Biology and Human Social Behavior*, ed. N.A. Cagnon and W.G. Irons, pp. 257–273. North Scituate, Mass.: Duxbury Press.

Irons, W.G. 1980. Is Yomut social behavior adaptive? In *Sociobiology: Beyond Nature/Nurture*, ed. G. Barlow and J. Silverber, pp. 417–463. Boulder, Col.: Westview Press.

Irons, W.G. 1983. Human female reproductive strategies. In *Social Behavior of Female Vertebrates*, ed. S.W. Wasser, pp. 169–213. New York: Academic Press.

Itani, J. 1959. Paternal care in the wild Japanese monkey, *Macaca fuscata fuscata*. *Primates* 2: 61–93.

Janson, C.H. 1984. Female choice and mating system of the brown capuchin monkey *Cebus apella*. *Zietschrift für Tierpsychologie* 65: 177–200.

Jay, P. 1965. The common langur of North India. In *Primate Behavior*, ed. I. DeVore, pp. 197–249. New York: Holt, Rinehart, and Winston.

Johanson, D.C., and M.A. Edey. 1981. *Lucy: The Beginnings of Humankind*. New York: Simon and Schuster.

Johnson, R.E. 1970. Some correlates of extramarital coitus. *Journal of Marriage and the Family* 32: 449–456.

Jolly, A. 1966. *Lemur Behavior*. Chicago: University of Chicago Press.

Jones, C.B. 1985. Reproductive patterns in mantled howler monkeys: Estrus, mate choice, and copulation. *Primates* 26: 130–142.

Kano, T. 1980. The social behavior of wild pygmy chimpanzees (*Pan paniscus*) of Wamba: A preliminary report. *Journal of Human Evolution* 9: 243–260.

Kano, T. 1992. *The Last Ape: Pygmy Chimpanzee Behavior and Ecology*. Palo Alto: Stanford University Press.

Kaplan, J.R. 1977. Patterns of fight interference in free-ranging rhesus monkeys. *American Journal of Physical Anthropology* 47: 279–288.

Kaufmann, J.H. 1965. A three-year study of mating behavior in a free-ranging band of rhesus monkeys. *Ecology* 46: 500–512.

Kawai, M. 1958. On the system of social ranks in a natural group of Japanese monkeys. *Primates* 1: 11–48.

Kawamura, S. 1958. Matriarchial social order in the Minoo-B group: A study on the rank system of Japanese macaques. *Primates* 1: 149–156.

Keddy, A.C. 1986. Female mate choice in vervet monkeys (*Cercopithecus aethiops*). *American Journal of Primatology* 10: 125–143.

Keddy, A.C., R.M. Seyfarth, and M.J. Raleigh. 1989. Male parental care, female choice, and the effect of an audience in vervet monkeys. *Animal Behaviour* 38: 262–271.

Kenny, J.A. 1973. Sexuality of pregnant and breastfeeding women. *Archives of Sexual Behavior* 2: 215–229.

Kinsey, A.C., W.B. Pomeroy, C.E. Martin, and P.H. Gebhard. 1953. *Sexual Behavior in the Human Female*. Philadelphia: W.B. Saunders.

Kirkpatrick, M. 1982. Sexual selection and the evolution of female choice. *Evolution* 39: 370–386.

Kirkpatrick, M. 1987. Sexual selection by female choice in polygynous animals. *Annual Review of Ecology and Systematics* 18: 43–70.

Klassen, A.D., C.J. Williams, and E.E. Levitt. 1989. *Sex and Morality in the U.S.* Middletown, Conn.: Wesleyan University Press.

Kodric-Brown, A., and J.H. Brown. 1984. Truth in advertising: The kinds of traits favored by sexual selection. *American Naturalist* 124: 309–323.

Konner, M. 1990. *Why the Reckless Survive and Other Secrets of Human Nature*. New York: Penguin Books.

Kuester, J., and A. Paul. 1984. Female reproductive characteristics in semi-free ranging Barbary macaques (*Macaca sylvanus*). *Folia Primatologica* 43: 69–83.

Kumar, A., and G. Kurup. 1985a. Intertroop interactions in the lion-tailed macaque. In *The Lion-Tailed Macaque: Status and Conservation*, ed. P.G. Heltne, pp. 91–107. New York: Alan R. Liss.

Kumar, A., and G. Kurup. 1985b. Sexual behavior of the lion-tailed macaque, *Macaca silenus*. In *The Lion-Tailed Macaque: Status and Conservation*, ed. P.G. Heltne, pp. 109–130. New York: Alan R. Liss.

Kummer, H. 1968. *Social Organization of Hamadryas Baboons*. Chicago: University of Chicago Press.

Kurland, J.A. 1977. Kin selection in the Japanese monkey. *Contribution to Primatology* 12: 1–145.

Lancaster, J.B. 1972. Play-mothering: The relations between juvenile females and young infants among free-ranging vervets. In *Primate Socialization*, ed. F.E. Poirier, pp. 83–104. New York: Random House.

Lancaster, J.B. 1979. Sex and gender in evolutionary perspective. In *Human Sexuality; A Comparative and Developmental Perspective*, ed. H. Katchadorian, pp. 51–80. Berkeley: University of California Press.

Lancaster, J.B. 1989. Women in biosocial perspective. In *Gender and Anthropology*, ed. S. Morgan, pp. 95–115. Washington, D.C.: American Anthropological Association.

Lancaster, J.B. 1991. A feminist and evolutionary biologist looks at women. *Yearbook of Physical Anthropology* 34: 1–11.

Lancaster, J.B., and H. Kaplan. 1992. Human mating and family formation strategies; The effects of variability among males in quality and the allocation of mating effort and parental investment. In *Topics in Primatology*, ed. T. Nishida, W.C. McGrew, P. Marler, M. Pickford and F.B. deWaal, pp. 21–33. Tokyo: University of Tokyo Press.

Lancaster, J., and C.S. Lancaster. 1983. Parental investment: The hominid adaptation. In *How Humans Adapt*, ed. D.J. Ortner, pp. 33–65. Washington, D.C.: Smithsonian Institution Press.

Lancaster, J.B., and C.S. Lancaster. 1987. The watershed: Change in parental-investment and family formation strategies in the course of human evolution. In *Parenting across the Lifespan: Biosocial Dimensions*, ed. J. B. Lancaster, J. Altmann, A.S. Rossi, and L.K. Sherrod, pp. 187–205. New York: Aldine de Gruyter.

Lande, R. 1981. Models of speciation by sexual selection on polygenic traits. *Proceedings of the National Academy of Sciences* 78: 3721–3725.

Lee, R.B. 1984. *The Dobe !Kung*, ed. G. Spindler and L. Spindler. New York: Holt, Rinehart, and Winston.

Lee, R.B., and I. DeVore. 1968. *Man the Hunter*. New York: Aldine.

Leibowitz, L. 1983. Origins of the sexual division of labor. In *Women's Nature: Rationalization of Inequality*, ed. M. Lowe and R. Hubbard, pp. 123–147. New York: Pergamon.

Leshner, A.I. 1978. *An Introduction to Behavioral Endocrinology*. New York: Oxford University Press.

Lightfoot-Klein, H. 1989. *Prisoners of Ritual: An Odyssey into Female Genital Circumcision in Africa*. Binghamton, N.Y.: Harrington Park Press.

Lindburg, D.G. 1969. Rhesus monkeys: Mating season mobility of adult males. *Science* 166: 1776–1778.

Lindburg, D.G. 1971. The rhesus monkey in northern India: An ecological and behavioral study. In *Primate Behavior*, ed. L.A. Rosenblum, pp. 1–106. New York: Academic Press.

Lindburg, D.G. 1980. *The Macaques: Studies in Ecology, Behavior, and Evolution*. New York: Van Nostrand Reinhold.

Lindburg, D.G. 1983. Mating behavior and estrus in the Indian rhesus monkey. In *Perspectives in Primate Biology*, ed. P.K. Seth, pp. 45–61. New Delhi: Today and Tomorrow's Publishers.

Lindburg, D.G. 1987. Seasonality of reproduction in primates. *Comparative Primate Biology* 2: 167–218.

Lindburg, D.G. 1990. Proceptive calling by female lion-tailed macaques. *Zoo Biology* 9: 437–446.

Lindburg, D.G., S. Shideler, and H. Fitch. 1985. Sexual behavior in relation to time of ovulation in the lion-tailed macaque. In *The Lion-Tailed Macaque: Status and Conservation*, ed. P.G. Heltne, pp. 131–148. New York: Alan R. Liss.

Linton, S. 1971. Woman the gatherer: Male bias in anthropology. In *Women in Perspective: A Guide for Cross-cultural studies*, ed. S.E. Jacobs, pp. 9–21. Urbana: University of Illinois Press.

Litchfield, H. 1915. *Emma Darwin: A Century of Family Letters, 1792–1896*. London: John Murrary.

Lovejoy, O.C. 1981. The origin of man. *Science* 211: 341–350.

Loy, J. 1971. Estrous behavior of free-ranging rhesus monkeys (*Macaca mulatta*). *Primates* 12: 1–31.

McClintock, M. 1971. Menstrual synchrony and suppression. *Nature* 229: 244–245.

MacFarland, L.Z. 1976. Comparative anatomy of the clitoris. In *The Clitoris*, ed. T.P. Lowry and T.S. Lowry, pp. 22–34. St. Louis: W.H. Green.

McKenna, J.J. 1979. The evolution of allomothering behavior among colobine monkeys: Function and opportunism in evolution. *American Anthropologist* 81: 818–840.

Majerus, M.E.N., P. O'Donald, P.W.E. Kearns, and H. Ireland. 1986. Genetics and the evolution of female choice. *Nature* 321: 164–167.

Manson, J.H. 1991. Female Mate Choice in Cayo Santiago Rhesus Macaques. Dis., University of Michigan.

Manson, J.H. 1992. Measuring female mate choice in Cayo Santiago rhesus macaques. *Animal Behaviour* 44: 405–416.

Margulis, L., and D. Sagan. 1985. *The Origins of Sex*. New Haven: Yale University Press.

Margulis, L., and D. Sagan. 1988. Sex: The canabalistic legacy of primordial androgynes. In *The Evolution of Sex*, ed. R. Bellig and G. Stevens, pp. 23–48. San Francisco: Harper and Row.

Martin, R.D., and A.F. Dixson. 1992. *Paternity in Primates: Genetic Tests and Theories*. Basel: Karger.

Mason, W. 1966. Social organization of the South American monkey *Callicebus moloch:* A preliminary report. *Tulane Studies in Zoology* 13: 23–28.

Masters, W., and V. Johnson. 1965a. The sexual response cycle of the human female. I. Gross anatomical considerations. In *Sex Research: New Developments*, ed. J. Money, pp. 53–89. New York: Holt, Rinehart, and Winston.

Masters, W., and V. Johnson. 1965b. The sexual response cycle of the human female. II. The clitoris: Anatomical and clinical considerations. In *Sex Research: New Developments*, ed. J. Money, pp. 90–104. New York: Holt, Rinehart, and Winston.

Masters, W. and V. Johnson. 1966. *Human Sexual Response*. Boston: Little, Brown.

Maynard Smith, J. 1955. Fertility, mating behavior, and sexual selection in *Drosophila subobscura*. *Genetics* 54: 261–279.

Maynard Smith, J. 1971. What use is sex? *Journal of Theoretical Biology* 30: 319–335.

Maynard Smith, J. 1978. *The Evolution of Sex*. Cambridge: Cambridge University Press.

Maynard Smith, J. 1991. Introduction. In *The Ant and the Peacock*, ed. H. Cronin, pp. ix–x. Cambridge: Press Syndicate of the University of Cambridge.

Melnick, D.J., M.C. Pearl, and A.F. Richard. 1984. Male migration and inbreeding avoidance in wild rhesus monkeys. 7: 229–243.

Michael, R.P., and E.B. Keverne. 1970. Primate sex pheromones of vaginal origin. *Nature* 225: 84–85.

Michael, R.P., E.B. Keverne, and R.W. Bonsall. 1971. Pheromones: Isolation of male sex attractants from a female primate. *Science* 172: 964–966.

Michael, R.P., and D. Zumpe. 1978. Potency in male rhesus monkeys: Effects of continuously receptive females. *Science* 200: 451–453.

Milton, K. 1985. Mating patterns of woolly spider monkeys, *Brachyteles arachnoides:* Implications for female choice. *Behavioral Ecology and Sociobiology* 17: 53–59.

Mitani, J.C. 1985a. Gibbon song duets and intergroup spacing behavior. *Behaviour* 92: 59–96.

Mitani, J.C. 1985b. Mating behaviour of male orangutans in the Kutai Game Reserve, Indonesia. *Animal Behaviour* 33: 392–402.

Moore, J., and R. Ali. 1984. Are dispersal and inbreeding avoidance related? *Animal Behaviour* 32: 94–112.

Moore, K.L. 1988. *The Developing Human: Clinically Oriented Embryology*. Philadelphia: W.B. Saunders.

Mori, U., and R.I.M. Dunbar. 1985. Changes in the reproductive condition of female gelada baboons following the take-over of one-male units. *Zeitschrift für Tierpsychologie* 67: 215–224.

Murdock, G.P., and D.R. White. 1969. Standard cross-cultural sample. *Ethnology* 8: 329–369.

Nicholson, N.A. 1987. Infants, mothers and other females. In *Primate Societies*, ed. B.B. Smuts, D.L. Cheney, R.M. Seyfarth, R. Wrangham and T.T. Strusaker, pp. 330–324. Chicago: University of Chicago Press.

Nishida, T. 1992. *The Chimpanzees of the Mahale Mountains: Sexual and Life History Strategies.* Tokyo: University of Tokyo Press.

Ober, C., T.J. Olivier, D.S. Sade, J.M. Schneider, J. Cheverud, and J. Buettner-Janusch. 1984. Demographic components of gene frequency change in free-ranging macaques on Cayo Santiago. *American Journal of Physical Anthropology* 64: 223–231.

O'Donald, P. 1980. *Genetic Models of Sexual Selection.* Cambridge: Cambridge University Press.

O'Donald, P. 1982. Fisher's contributions to the theory of sexual selection as the basis of recent research. *Theoretical Population Biology* 96: 281–294.

Olson, D.K. 1985. The importance of female choice in the mating system of wild patas monkeys. *American Journal of Physical Anthropology* 66: 211.

Overstreet, J., and D.F. Katz. 1977. Sperm transport and selection in the female genital tract. In *Development in Mammals*, ed. M.H. Johnson, pp. 31–65. Amsterdam: North Holland.

Palombit, R. 1992. Pair bonds and monogamy in wild siamang (*Hylobates syndactylus*) and white-handed gibbons (*Hylobates lar*) in northern Sumatra. Ph.D. diss., University of California, Davis.

Parker, G.A. 1982. Why are there so many tiny sperm? *Journal of Theoretical Biology* 96: 281–294.

Parker, G.A. 1984. Sperm competition. In *Sperm Competition and Animal Mating Systems*, ed. R.L. Smith New York: Academic Press.

Partridge, L. 1980. Mate choice increases a component of offspring fitness in fruit flies. *Nature* 283: 290–291.

Pereira, M.E., and M.E. Weiss. 1991. Female mate choice, male migration, and the threat of infanticide in ringtail lemurs. *Behavioral Ecology and Sociobiology* 28: 141–152.

Prentice, A., and A. Prentice. 1988. Reproduction against the odds. *New Scientist* 118: 42–46.

Price, E.C. 1991. Infant carrying as a courtship strategy of breeding male cotton-top tamarins. *Animal Behaviour* 40: 784–786.

Quiatt, D. 1979. Aunts and mothers: Adaptive implications of allomaternal behavior of nonhuman primates. *American Anthropologist* 81: 310–319.

Ransom, T.W. 1981. *Beach Troop at Gombe*. Lewisburg, Pa.: Bucknell University Press.

Rasmussen, K.L.R. 1983. Influences of affiliative preferences upon behavior of male and female baboons during sexual consprtships. In *Primate Social Relationships: An Integrated Approach*, ed. R.A. Hinde, pp. 116–120. Oxford: Blackwell.

Rawlins, R.G., and M.J. Kessler. 1986. *The Cayo Santiago Macaques: History, Behavior, and Biology*. Albany: State University of New York Press.

Richard, A.F. 1987. Malagasy prosimians: Female dominance. In *Primate Societies*, ed. B.B. Smuts, D.L. Cheney, R.M. Seyfarth, R. Wrangham and T.T. Strusaker, pp. 25–33. Chicago: University of Chicago Press.

Richards, E. 1983. Darwin and the descent of women. In *The Wider Domain of Evolutionary Thought*, ed. D. Oldroyd and I. Langham, pp. 57–111. New York: Reidel.

Ripley, S. 1980. Infanticide in langurs and man: Adaptive advantages or social pathology. In *Biosocial Mechanisms of Population Regulation*, ed. R. Tuttle, pp. 349–390. New Haven: Yale University Press.

Robinson, J.G. 1982. Intrasexual competition and mate choice in primates. *American Journal of Primatology* 1 (Supplement): 131–144.

Robinson, J.G., and C.H. Janson. 1987. Capuchins, squirrel monkeys, and atelines: Socioecological convergence with Old World primates. In *Primate Societies*, ed. B.B. Smuts, D.L. Cheney, R.M. Seyfarth, R. Wrangham, and T.T. Strusaker, pp. 69–82. Chicago: University of Chicago Press.

Rodman, P.S., and J.C. Mitani. 1987. Orangutans: Sexual dimorphism in a solitary species. In *Primate Societies*, ed. B.B. Smuts, D.L. Cheney, R.M. Seyfarth, R. Wrangham and T.T. Strusaker, pp. 146–154. Chicago: University of Chicago Press.

Roonwal, M.L., and S.M. Mohnot. 1977. *Primates of South Asia: Ecology, Sociobiology, and Behavior*. Cambridge: Harvard University Press.

Rosenblatt, P.C., and R.M. Anderson. 1981. Human sexuality in cross-cultural perspective. In *The Basis of Human Sexual Attraction*, ed. M. Cook, pp. 215–250. New York: Academic Press.

Rowell, T.E. 1966. Forest living baboons in Uganda. *Journal of the Zoological Society of London* 147: 344–364.

Rowell, T.E. 1967. A quantitative comparison of the behavior of a wild and caged baboon troop. *Animal Behaviour* 15: 499–509.

Russell, M.J., G.M. Switz, and K. Thompson. 1980. Olfactory influences on the human menstrual cycle. *Pharmacology, Biochemistry, and Behavior* 13: 737–738.

Ryan, M.J. 1983. Sexual selection and communication in a neotropical frog, *Physalaemus pustulosus*. *Evolution* 37: 261–271.

Ryan, M.J. 1990. Signals, species, and selection. *American Scientist* 78: 46–52.

Saayman, G.S. 1970. The menstrual cycle and sexual behavior in a troop of free-ranging chacma baboons (*Papio ursinus*). *Folia Primatologica* 12: 81–110.

Sade, D.S. 1965. Some aspects of parent-offspring and sibling relationships in a group of rhesus monkeys, with a discussion of grooming. *American Journal of Physical Anthropology* 23: 1–18.

Sarich, V.M., and A.C. Wilson. 1967. Immunological time scale for hominid evolution. *Science* 158: 1200–1203.

Sauther, M.L. 1991. Reproductive behavior of free-ranging *Lemur catta* at Beza Mahafaly Special Reserve, Madagascar. *American Journal of Physical Anthropology* 84: 463–477.

Scott, L. 1984. Reproductive behavior of adolescent female baboons (*Papio anubis*) in Kenya. In *Female Primates: Studies by Women Primatologists*, ed. M.F. Small, pp. 77–100. New York: Alan R. Liss.

Seger, J. 1985. Unifying genetic models for the evolution of female choice. *Evolution* 39: 1185–1193.

Seyfarth, R.M. 1978. Social relationships among adult male and female baboons, II. Behaviours throughout the female reproductive cycle. *Behaviour* 64: 227–247.

Sherfey, M.J. 1966. *The Nature and Evolution of Female Sexuality.* New York: Random House.

Short, R.V. 1976. The evolution of human reproduction. *Proceedings of the Royal Society of London* 195: 3–24.

Short, R.V. 1979. Sexual selection and it component parts, somatic and genetical selection as illustrated by man and the great apes. *Advances in the Study of Behavior* 9: 131–158.

Silk, J.B. 1980. Kidnapping and female competition among captive bonnet macaques. *Primates* 21: 100–110.

Silk, J.B. 1988. Social mechanisms of population regulation in a captive group of bonnet macaques (*Macaca radiata*). *American Journal of Primatology* 14: 111–124.

Silk, J.B., and R. Boyd. 1983. Cooperation, competition, and mate choice in matrilineal macaque groups. In *Social Behavior of Female Vertebrates*, ed. S.K. Wasser, pp. 315–347. New York: Academic Press.

Silk, J.B., A. Samuels, and P.S. Rodman. 1981. The influences of kinship, rank, and sex on affiliation and aggression between adult female and immature bonnet macaques (*Macaca radiata*). *Primates* 78: 111–117.

Sleeth Mosedale, S. 1978. Science corrupted: Victorian biologists consider "the woman question." *Journal of the History of Biology* 11: 1–58.

Small, M.F. 1981. Body fat, rank, and nutritional status in a captive group of rhesus macaques. *International Journal of Primatology* 2: 91–96.

Small, M.F. 1982. A comparison of mother and nonmother behavior during birth season in two species of captive macaques. *Folial Primatologica* 38: 99–107.

Small, M.F. 1984a. Aging and reproductive success in female *Macaca mulatta*. In *Female Primates: Studies by Women Primatologists*, ed. M.F. Small, pp. 249–259. New York: Alan R. Liss.

Small, M.F., ed. 1984b. *Female Primates: Studies by Women Primatologists*. New York: Alan R. Liss.

Small, M.F. 1988. Female primate sexual behavior and conception: Are there really sperm to spare? *Current Anthropology* 29: 81–100.

Small, M.F. 1989. Female choice in nonhuman primates. *Yearbook of Physical Anthropology* 32: 103–127.

Small, M.F. 1990a. Political animal. *The Sciences* March/April: 36–55.

Small, M.F. 1990b. Promiscuity in Barbary macaques (*Macaca sylvanus*). *American Journal of Primatology* 20: 267–282.

Small, M.F. 1990c. Consortships and conceptions in captive rhesus macaques (*Macaca mulatta*). *Primates* 31: 339–350.

Small, M.F. 1990d. Alloparental behavior in Barbary macaques (*Macaca sylvanus*). *Animal Behaviour* 39: 297–306.

Small, M.F. 1990e. Social climber: Independent rise in rank by a female Barbary macaque (*Macaca sylvanus*). *Folia Primatologica* 55: 85–91.

Small, M.F. 1992a. The evolution of female sexuality and mate selection in humans. *Journal of Human Nature* 3: 133–156.

Small, M.F. 1992b. Female choice in mating. *American Scientist* 80: 142–151.

Small, M.F. 1992c. What's love got to do with it? *Discover* 13: 46–51.

Small, M.F. 1993. Closing the gap. *Wildlife Conservation*. In press.

Small, M.F. Ovulation revealed: a pilot study of college women and men. Unpublished manuscript.

Small, M., and R. Palombit. Female choice is not always sexual selection. Unpublished manuscript.

Small, M.F., and D.G. Smith. 1982. The relationship between maternal and paternal rank in rhesus macaques (*Macaca mulatta*). *Animal Behaviour* 30: 626–627.

Smith, D.G. 1982. Use of genetic markers in the colony management of nonhuman primates: A review. *Laboratory Animal Science* 32: 540–546.

Smith, D.G., and M.F. Small. 1987. Mate choice by lineage in three captive groups of rhesus macaques (*Macaca mulatta*). *American Journal of Physical Anthropology* 73: 185–191.

Smith, T.W. 1990. The sexual revolution? *Public Opinion Quarterly* 54: 415–435.

Smith, T.W. 1991. Adult sexual behavior in 1989: Number of partners, frequency of intercourse and risk of AIDS. *Family Planning Perspectives* 23: 102–107.

Smuts, B.B. 1985. *Sex and Friendship in Baboons*. New York: Aldine.

Smuts, B.B. 1992. Male aggression against women: An evolutionary perspective. *Journal of Human Nature* 3: 1–44.

Smuts, B.B., and R.W. Smuts. Male aggression and sexual coersion of females in nonhuman primates and other mammals: Evidence and theoretical implications. In *Advances in the Study of Behavior*, ed. P.B. Slater. New York: Academic Press. In press.

Sommer, V. 1988. Female-female mounting in langurs. *International Journal of Primatology* 8: 478.

Sommer, V. 1989. Sexual harassment in langur monkeys (*Presbytis entellus*). *Ethology* 80: 205–217.

Sommer, V., A. Srivastava, and C. Borries. Cycles, sexuality, and conception in free-ranging langurs (*Presbytis entellus*). *American Journal of Primatology*. In press.

Southwick, C.H., M.A. Beg, and M.R. Siddiqi. 1965. Population dynamics of rhesus monkeys in Northern India. In *Primate Behavior*, ed. I. DeVore, pp. 111–159. New York: Holt, Rinehart, and Winston.

Southwick, C.H., and R.B. Smith. 1986. The growth of primate field studies. In *Comparative Primate Biology, Vol. 2A: Behavior, Conservation, and Ecology*, ed. G. Mitchell and J. Erwin, pp. 73–91. New York: Alan R. Liss.

Srivastava, A., C. Borries, and V. Sommer. 1991. Homosexual mounting in free-ranging Hanuman langurs. *Archives of Sexual Behavior* 20: 487–516.

Stacey, P.B. 1982. Female promiscuity and male reproductive success in social birds and mammals. *American Naturalist* 120: 51–64.

Stern, B.R, and D.G. Smith. 1984. Sexual behavior and paternity in three captive groups of rhesus monkeys (*Macaca mulatta*). *Animal Behaviour* 32: 23–32.

Stewart, K.J., and A.H. Harcourt. 1987. Gorillas: Variation in female relationships. In *Primate Societies*, ed. B.B. Smuts, D.L. Cheney, R.M. Seyfarth, R. Wrangham, and T.T. Strusaker, pp. 155–164. Chicago: University of Chicago Press.

Strassmann, B.I. 1981. Sexual selection, paternal care, and concealed ovulation in humans. *Ethology and Sociobiology* 2: 31–40.

Strier, K.B. 1990. New World primates, new frontiers: Insights from the woolly spider monkey, or muriqui (*Brachyteles arachnoides*). *International Journal of Primatology* 11: 7–19.

Strier, K.B. 1992. Causes and consequences of nonaggression in the woolly spider monkey, or muriqui (*Brachyteles arachnoides*). In *Aggression and*

Peacefulness in Humans and Other Primates, ed. J. Silverberg and P. Gray, pp. 100–116. Oxford: Oxford University Press.

Strum, S.C. 1975. Life with the Pumphouse gang: New insights into baboon behavior. *National Geographic* 147: 672–691.

Strum, S.C. 1982. Agonistic dominance in male baboon: An alternate view. *International Journal of Primatology* 3: 175–207.

Susman, R. L. 1984. *The Pygmy Chimpanzee: Evolution, Biology, and Behavior*. New York: Plenum.

Symons, D. 1979. *The Evolution of Human Sexuality*. Oxford: Oxford University Press.

Takahata, Y. 1982a. Social relations between adult male and female Japanese monkeys in Arashiyama B troop. *Primates* 23: 1–23.

Takahata, Y. 1982b. The socio-sexual behavior of Japanese monkeys. *Zeitschrift für Tierpsychologie* 59: 89–104.

Tanner, N. 1981. *On Becoming Human*. Cambridge: Cambridge University Press.

Taub, D.M. 1980a. Female choice and mating strategies among wild Barbary macaques (*Macaca sylvanus*). In *The Macaques*, ed. D. L. Lindburg, pp. 287–344. New York: Van Nostrand Reinhold.

Taub, D.M. 1980b. Testing the "agonistic buffering" hypothesis, 1: The dynamics of participation in the triadic interaction. *Behavioral Ecology and Sociobiology* 6: 187–197.

Taub, D.M. 1984. *Primate Paternalism*. New York: Van Nostrand Reinhold.

Tavris, C., and S. Sadd. 1975. *The Redbook Report on Female Sexuality*. New York: Delacorte.

Terborgh, J., and A. Wilson Goldizen. 1985. On the mating system of the cooperatively breeding saddle-backed tamarin (*Saguinus fuscicollis*). *Behavioral Ecology and Sociobiology* 16: 293–299.

Thompson-Handler, N, R. Malenky, and N. Badrian. 1984. Sexual behavior of *Pan paniscus* under natural condition in the Lomako Forest, Equateur, Zaire. In *The Pygmy Chimpanzee: Evolution, Biology, and Behavior*, ed. R.L. Susman, pp. 347–368. New York: Plenum Press.

Trivers, R.L. 1971. The evolution of reciprocal altruism. *Quarterly Review of Biology* 46: 35–57.

Trivers, R.L. 1972. Parental investment and sexual selection. In *Sexual Selection and the Descent of Man*, ed. B. Campbell, pp. 1136–1179. Chicago: Aldine.

Tsingalia, H.M., and T.E. Rowell. 1984. The behavior of adult male blue monkeys. *Zietschrif fur Tierpsychologie* 64: 253–268.

Turke, P.W. 1984. Effects of ovulatory concealment and synchrony of protohominid mating systems and parental roles. *Ethology and Sociobiology* 5: 33–44.

Turke, P.W., and L.L. Betzig. 1985. Those who can do: Wealth, status, and reproduction on Ifaluk. *Ethology and Sociobiology* 6: 79–87.

Tutin, C.E. 1979. Mating patterns and reproductive strategies in a community of wild chimpanzees (*Pan troglodytes schweinfurthii*). *Behavioral Ecology and Sociobiology* 6: 29–38.

Tutin, C., and P. McGinnis. 1981. Chimpanzee reproduction in the wild. In *Reproductive Biology of the Great Apes*, ed. C.E. Graham, pp. 239–264. New York: Academic Press.

Udry, J.B., N.M. Morris, and L. Walker. 1973. Effect of contraceptive pills on sexual activity in the luteal phase of the human menstrual cycle. *Archives of Sexual Behavior* 2: 205–214.

van den Bergh, U., and J. Lehmann. 1991. Signal exchange among Barbary macaques during mating season. *Primate Report* 31: 37.

van Noordwijk, M.A. 1985. Sexual behavior of Sumatran long-tailed macaques (*Macaca fascicularis*). *Zeitschrift für Tierpsychologie* 70: 277–296.

Veith, J.L., M. Buck, S. Getzlaf, P. Van Dalfsen, and S. Slade. 1983. Exposure to men influences the occurrence of ovulation in women. *Physiology and Behavior* 31: 313–315.

Vining, D.R. 1986. Social versus reproductive success: The central theoretic problem of human sociobiology. *Behavioral and Brain Sciences* 9: 167–216.

von Schantz, T., G. Göransson, G. Andersson, I. Fröberg, M. Grahn, H. Helgée, and W. Håkan. 1989. Female choice selects for viability-based male trait in pheasants. *Nature* 337: 166–169.

Wallis, S.J. 1983. Sexual behavior and reproduction of *Cercocebus albigena johnstonii* in Kibale Forest, Western Uganda. *International Journal of Primatology* 4: 153–166.

Walters, J.R. 1987. Transition to adulthood. In *Primate Societies*, ed. B.B. Smuts, D.L. Cheney, R.M. Seyfarth, R. Wrangham and T.T. Strusaker, pp. 358–369. Chicago: University of Chicago Press.

Waser, P. 1978. Postreproductive survival and behavior in a free-ranging female mangabey. *Folia Primatologica* 29: 142–160.

Wasser, S.K., and M.L. Waterhouse. 1983. The establishment and maintenance of sex biases. In *Social Behavior of Female Vertebrates*, ed. S. K. Wasser, pp. 19–35. New York: Academic Press.

Watts, D.P. 1991. Mountain gorilla reproduction and sexual behavior. *American Journal of Primatology* 24: 211–255.

Weismann, A. 1887. *The Significance of Sexual Reproduction in the Theory of Natural Selection*. Oxford: Oxford University Press.

White, F.J., and N. Thompson-Handler. Ecological and social correlates of female homosexual behavior in *Pan paniscus*. In press.

Whitten, P.L. 1983. Diet and dominance among female vervet monkeys (*Cercopithecus aethiops*). *American Journal of Primatology* 5: 139–159.

Whittenberger, J.F. 1981. *Animal Social Behavior*. Boston: Duxbury Press.

Whyte, M.K. 1978. Cross-cultural codes dealing with the relative status of women. *Ethnology* 17: 211–237.

Wildt, D.E., L.L. Doyle, S.C. Stone, and R.M. Harrison. 1977. Correlation of perineal swelling with serum ovarian hormone levels, vaginal cytology, and ovarian follicular development during the baboon reproductive cycle. *Primates* 18: 261–270.

Williams, G.C. 1975. *Sex a. 1 Evolution.* Monographs in Population Biology. Princeton: Princeton University Press.

Wilson, E.O. 1975. *Sociobiology: The New Synthesis.* Cambridge: Harvard University Press.

Wilson Goldizen, A. 1987. Tamarins and marmosets: Communal care of offspring. In *Primate Societies*, ed. B.B. Smuts, D.L. Cheney, R.M. Seyfarth, R. Wrangham and T.T. Strusaker, pp. 34–43. Chicago: University of Chicago Press.

Wolfe, L. 1980. The sexual profile of the Cosmopolitan girl. *Cosmopolitan* September: 254–265.

Wolfe, L.M. 1979. Behavioral patters of estrous females of the Arashiyama West troop of Japanese macaques (*Macaca fuscata*). *Primates* 20: 525–534.

Wolfe, L.M. 1984. Female rank and reproductive success among Arashiyama B Japanese macaques (*Macaca fuscata*). *International Journal of Primatology* 5: 133–143.

Wolfe, L.M. 1986. Sexual strategies of female Japanese macaques (*Macaca fuscata*). *Human Evolution* 1: 267–275.

Wood, F.W. 1990. *An American profile—opinions and behavior, 1972–1989.* Detroit: Gale Research.

Wrangham, R.W. 1980. An ecological model of female-bonded primate groups. *Behaviour* 75: 262–300.

Wright, P.C. 1978. Home range, activity patterns, and agonistic encounters of a group of night monkeys (*Aotus trivergatus*) in Peru. *Folia Primatologica* 29: 43–55.

Yamada, M. 1963. A study of blood-relationships in the natural society of Japanese macaques: An analysis of co-feeding, grooming, and playmate relationships in Minoo-B troop. *Primates* 4: 43–65.

Yamagiwa, J. 1985. Socio-sexual factors of troop fission in wild Japanese monkeys (*Macaca fuscata yakiu*) on Yakushima Island, Japan. *Primates* 26: 105–120.

Zahavi, A. 1975. Mate selection—a selection for a handicap. *Journal of Theoretical Biology* 53: 205–214.

Zihlman, A.L., and N. Tanner. 1978. Gathering and the hominid adaptation. In *Female Hierarchies*, ed. L. Tiger and T. Fowler, pp. 163–194. Chicago: Beresford.

Zumpe, D., and R.P. Michael. 1968. The clutching reaction and orgasm in the female rhesus monkey (*Macaca mulatta*). *Journal of Endocrinology* 40: 117–123.

Index